WITHDRAWN BY THE
UNIVERSITY OF MICHIGAN

STUDIES IN COMBINATORICS

MAA STUDIES IN MATHEMATICS

Published by
THE MATHEMATICAL ASSOCIATION OF AMERICA

Committee on Publications
E. F. BECKENBACH, Chairman

Subcommittee on MAA Studies in Mathematics
G. L. WEISS, Chairman
F. J. ALMGREN, JR.
WANDA W. HELM
A. C. TUCKER

Studies in Mathematics

Volume 1: STUDIES IN MODERN ANALYSIS
edited by R. C. Buck

Volume 2: STUDIES IN MODERN ALGEBRA
edited by A. A. Albert

Volume 3: STUDIES IN REAL AND COMPLEX ANALYSIS
edited by I. I. Hirschman, Jr.

Volume 4: STUDIES IN GLOBAL GEOMETRY AND ANALYSIS
edited by S. S. Chern

Volume 5: STUDIES IN MODERN TOPOLOGY
edited by P. J. Hilton

Volume 6: STUDIES IN NUMBER THEORY
edited by W. J. LeVeque

Volume 7: STUDIES IN APPLIED MATHEMATICS
edited by A. H. Taub

Volume 8: STUDIES IN MODEL THEORY
edited by M. D. Morley

Volume 9: STUDIES IN ALGEBRAIC LOGIC
edited by Aubert Daigneault

Volume 10: STUDIES IN OPTIMIZATION
edited by G. B. Dantzig and B. C. Eaves

Volume 11: STUDIES IN GRAPH THEORY, PART I
edited by D. R. Fulkerson

Volume 12: STUDIES IN GRAPH THEORY, PART II
edited by D. R. Fulkerson

Volume 13: STUDIES IN HARMONIC ANALYSIS
edited by J. M. Ash

Volume 14: STUDIES IN ORDINARY DIFFERENTIAL EQUATIONS
edited by Jack Hale

Volume 15: STUDIES IN MATHEMATICAL BIOLOGY, PART I
edited by S. A. Levin

Volume 16: STUDIES IN MATHEMATICAL BIOLOGY, PART II
edited by S. A. Levin

Volume 17: STUDIES IN COMBINATORICS
edited by G.-C. Rota

Volume 18: STUDIES IN PROBABILITY THEORY
edited by Murray Rosenblatt

Volume 19: STUDIES IN STATISTICS
edited by R. V. Hogg

Volume 20: STUDIES IN ALGEBRAIC GEOMETRY
edited by A. Seidenberg

Volume 21: STUDIES IN FUNCTIONAL ANALYSIS
edited by R. G. Bartle

Volume 22: STUDIES IN COMPUTER SCIENCE
edited by S. V. Pollack

Volume 23: STUDIES IN PARTIAL DIFFERENTIAL EQUATIONS
edited by Walter Littman

Volume 24: STUDIES IN NUMERICAL ANALYSIS
edited by Gene H. Golub

Volume 25: STUDIES IN MATHEMATICAL ECONOMICS,
edited by Stanley Reiter

Tom Brylawski
University of North Carolina

R. L. Graham
Bell Telephone Laboratories

Curtis Greene
University of Pennsylvania

Marshall Hall, Jr.
California Institute of Technology

Douglas G. Kelly
University of North Carolina

Daniel J. Kleitman
Massachusetts Institute of Technology

Gian-Carlo Rota
Massachusetts Institute of Technology

B. L. Rothschild
University of California, Los Angeles

H. J. Ryser
California Institute of Technology

Joel Spencer
State University of New York, Stony Brook

Richard P. Stanley
Massachusetts Institute of Technology

Studies in Mathematics

Volume 17

STUDIES IN COMBINATORICS

Gian-Carlo Rota, editor

Massachusetts Institute of Technology

Published and distributed by
The Mathematical Association of America

MATH
QA
164
.S841

© *1978 by*
The Mathematical Association of America (Incorporated)
Library of Congress Catalog Card Number 78-60730

Complete Set ISBN 0-88385-100-8
Vol. 17 ISBN 0-88385-117-2

Printed in the United States of America

Current printing (last digit):

10 9 8 7 6 5 4 3 2

INTRODUCTION

Combinatorics has recently awakened from a long slumber, which began at the time of Euler. It has emerged as a new subject standing at the crossroads between pure and applied mathematics, the center of bustling activity, a simmering pot of new problems and exciting speculations.

The seven papers of this survey represent a wide enough sampling of current trends, from which the reader may, at least, extrapolate some of the missing material. They bear in common the one characteristic of contemporary combinatorics: striving for general new results, while using old and new problems as a test of efficiency.

The introduction to matroid theory by Brylawski and Kelly describes a theory whose background—at least from a distance—is the four-color conjecture, much like the background of algebraic number theory was, at least at the beginning, Fermat's conjecture. It matters little that neither theory has succeeded in solving its motivating problem. Too early a solution might, in fact, have prevented useful and unexpected applications. The theory of matroids is in some ways a generalization of the theory of electrical circuits. Its founder, Hassler Whitney, noticed the strange similarity between certain patterns of reasoning used in circuit theory and similar patterns in linear algebra, centered around the use of the Mac Lane-Steinitz exchange axiom. After him, it was believed for a while that matroids, namely, closure relations enjoying the Mac Lane-Steinitz exchange property, would sooner or later be reduced to their more concrete and better understood examples, namely, sets of points in projective space. This is the problem of representation or coordinatization of matroids, that has led to brilliant partial solutions by Tutte and others. However, the picture that is emerging from years of intensive study is the opposite. Most matroids are not representable in vector spaces at all. It is linear algebra that is, in a sense not yet fully understood, a special case of the theory of matroids; it is matroids that are contributing new concepts to linear algebra, in the wake of circuit theory.

Matroid theory, also called combinatorial geometry, led to the first step towards a combinatorial analog of K-theory, namely Brylawski's Tutte-Grothendieck ring. The critical problem generalizes to all matroids the problem of coloring a planar map. Whether the four-color conjecture is solved or not matters little as long as the reason for its difficulty is not understood, for such an understanding will raise it above the level of a silly curiosity and lead to new combinatorial techniques of general interest. The critical problem specializes by selecting suitable classes of matroids into a variety of problems of decreasing difficulty, all of them bearing a recognizable affinity to the coloring problem. It appears at present to be the most likely line for a general attack on this mysterious corner of mathematics.

If asked to name one and only one elegant theorem in their field, most combinatorialists would name Ramsey's theorem. For a long time it remained an isolated result, a rival to van der Waerden's theorem, which year after year would find startling new applications. Graham and Rothschild, later followed by Leeb and others, began to generalize the underlying pigeonhole reasoning to more complex situations, such as vector spaces over finite fields, and eventually discovered the natural setup, which can be elegantly described in terms of category theory. The resulting proofs are simple and streamlined, and "Pascal theory", as it has been named by Leeb, has yielded a bounty of applications with no end in sight.

Ever since Erdös marched onto the scene, the proof of the combinatorial pudding has been ext :mal set theory. The uncanny proofs, the unexpectedness of the conclusions were to become even more mysterious after Kleitman and several others joined in, and for a while any attempt at general understanding seemed hopeless.

The first crack was the inequality proved almost simultaneously by Lubell, Meschalkin and Yamamoto, which pointed to a "difference calculus" on partially ordered sets as a possible line of attack. Greene and Kleitman give a thorough exposition of this line of reasoning, which promises a bountiful crop of new results.

Block designs are generally acknowledged to be the most complex mathematical structure that can be defined from scratch in a few lines. Progress in understanding and classification has been slow and has proceeded by leaps and bounds, one ray of sunlight being followed by years of darkness. Largely through the efforts of Bose, Marshall Hall, and more recently of Ray-Chaudhuri and Wilson, this field has been enriched and made even more mysterious, a battleground of number-theory, projective geometry and plain cleverness. This is probably the most difficult combinatorics going on today, as Marshall Hall's survey shows.

The interaction between matrix theory and combinatorics began early in the game, when nineteenth-century invariant theory was in full swing. Petersen's work on graph theory was motivated by a problem of Hilbert, and Frobenius's definition of the free matrix of an incidence relation was suggested by his work on positive operators. Ryser's survey stresses the feeling now held by many that this may prove the unifying concept for matching theory, contrary to Frobenius's opinion. Jurkat and Ryser's amazing factorization of determinants and permanents is a gold mine that awaits exploitation. It may be expected that, with the current revival of concrete invariant theory, this field will enjoy a burst of renewed activity.

The application of probabilistic methods to combinatorics is another idea of Erdös that has borne fruit. Simple facts about finite sets and graphs, estimates of Ramsey numbers, threshold estimates on the growth of combinatorial structures, can at present only be established by probabilistic estimates. The reason for the success of the method lies probably in the fact that working with random variables instead of probabilities is a simple way of linearizing non-linear situations. One suspects a yet-to-be-brought-out new logic behind the method, as Spencer suggests in his survey.

One way to solve a problem in combinatorics is to transform it into a problem of algebra, and then turn it over to the algebraists.

Following this line, combinatorics has become the source of new algebraic structures—where else could algebraists get new structures from? It has also provided new examples and applications of old structures, like commutative rings in Stanley's work on polyhedra.

Perhaps the simplest way of making algebra out of combinatorics is the generating function, a concept that, starting from humble beginnings, is now, after suitable generalization, threatening to become a new way of combinatorial life. Stanley's survey displays some of the tantalizing connections of combinatorial structures with the theory of algebraic functions and with commutative algebra generally, and like all other papers of this survey, opens a horizon of possibilities.

GIAN-CARLO ROTA

CONTENTS

INTRODUCTION
Gian-Carlo Rota, vii

COMBINATORIAL MATRIX THEORY
H. J. Ryser, 1

PROOF TECHNIQUES IN THE THEORY OF FINITE SETS
Curtis Greene and Daniel J. Kleitman, 22

RAMSEY THEORY
R. L. Graham and B. L. Rothschild, 80

GENERATING FUNCTIONS
Richard P. Stanley, 100

NONCONSTRUCTIVE METHODS IN DISCRETE MATHEMATICS
Joel Spencer, 142

MATROIDS AND COMBINATORIAL GEOMETRIES
Tom Brylawski and Douglas G. Kelly, 179

COMBINATORIAL CONSTRUCTIONS
Marshall Hall, Jr., 218

INDEX, 255

COMBINATORIAL MATRIX THEORY[†]

H. J. Ryser

1. INTRODUCTION

We begin with an intriguing example that shows how a matrix equation may be used to represent a certain combinatorial problem. Suppose that we have a rectangle R of integral height m and of integral length n and that we partition all of R into t smaller rectangles. Each of these smaller rectangles is also required to have integral height and integral length. We then number the smaller rectangles that partition R in an arbitrary manner $1, \ldots, t$. A simple example with $m = 4$, $n = 5$, and $t = 6$ is illustrated in Figure 1.

Let A be a matrix of m rows and n columns. We say that A is of *size m by n*. Suppose that the elements of A are the integers 0 and

---

[†] This article is the outgrowth of the author's lectures presented to the Conference on Theoretical Matrix Theory at the University of California, Santa Barbara, December 10-14, 1973. The research was supported in part by grants from the Army Research Office-Durham and the National Science Foundation.

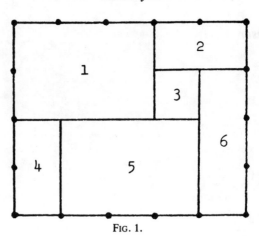

Fig. 1.

1. Then we call A a $(0,1)$-*matrix*. We now associate with the rectangles of Figure 1 the following two $(0,1)$-matrices of sizes 4 by 6 and 6 by 5, respectively:

$$X = \begin{bmatrix} 1 & 1 & 0 & 0 & 0 & 0 \\ 1 & 0 & 1 & 0 & 0 & 1 \\ 0 & 0 & 0 & 1 & 1 & 1 \\ 0 & 0 & 0 & 1 & 1 & 1 \end{bmatrix}, \quad Y = \begin{bmatrix} 1 & 1 & 1 & 0 & 0 \\ 0 & 0 & 0 & 1 & 1 \\ 0 & 0 & 0 & 1 & 0 \\ 1 & 0 & 0 & 0 & 0 \\ 0 & 1 & 1 & 1 & 0 \\ 0 & 0 & 0 & 0 & 1 \end{bmatrix}. \quad (1.1)$$

The number of 1's in column i of X is equal to the height of rectangle i and all of the 1's in column i of X occur consecutively. The topmost and bottommost 1's in column i of X locate the position of rectangle i with respect to the top and bottom horizontal lines of the full rectangle. Similarly, the number of 1's in row j of Y is equal to the length of rectangle j and all of the 1's in row j of Y occur consecutively. The first and last 1's in row j of Y locate the position of rectangle j with respect to the left and right vertical lines of the full rectangle. Suppose now that we multiply the matrices X and Y together. Then we obtain

$$XY = J, \qquad (1.2)$$

where J is the matrix of 1's of size 4 by 5.

Now this state of affairs is no mere accident. In fact it follows easily from the definition of matrix multiplication that the partitioning of rectangles of the type described is precisely equivalent to the study of the matrix equation (1.2), where X and Y are (0,1)-matrices of sizes m by t and t by n, respectively, such that the 1's in the columns of X and the rows of Y occur consecutively, and J is the matrix of 1's of size m by n.

The theory of matrices provides an extremely powerful tool for the study of a wide variety of combinatorial problems. In what follows we survey a number of diversified topics. The entire area is an exceedingly active one and we discuss both classical and recent results.

2. THE INCIDENCE MATRIX

Let $X = \{x_1, \ldots, x_n\}$ be a nonempty set of n elements. We call X an *n-set*. Now let $X_1, \ldots, X_m$ be m not necessarily distinct subsets of the n-set X. Such configurations are of great generality and occur repeatedly throughout the combinatorial literature. We set $a_{ij} = 1$ if $x_j \in X_i$ and we set $a_{ij} = 0$ if $x_j \notin X_i$. The resulting (0,1)-matrix

$$A = [a_{ij}] \qquad (i = 1, \ldots, m; \; j = 1, \ldots, n) \qquad (2.1)$$

of size m by n is the *incidence matrix* for the subsets $X_1, \ldots, X_m$ of X. Row i of A displays the subset X_i and column j of A displays the occurrences of the element x_j among the subsets. Thus A gives us a complete description of the subsets and the occurrences of the elements within the subsets.

We may relabel the subsets and the elements of the n-set and this does not destroy the combinatorial content of our original configuration. In terms of incidence matrices this means that we have replaced our original incidence matrix A by a new incidence matrix of the form

$$PAQ, \tag{2.2}$$

where P is a permutation matrix of order m and Q is a permutation matrix of order n. It follows that we are primarily concerned with properties of the incidence matrix that remain invariant under arbitrary permutations of rows and columns.

The representation of m subsets of an n-set by way of a (0,1)-matrix A of size m by n is of the utmost importance because it allows us to employ the powerful techniques of matrix theory to the study of combinatorial configurations. We may always form the matrix equations

$$AA^T = B \tag{2.3}$$

and

$$A^T A = C, \tag{2.4}$$

where A^T denotes the transpose of the matrix A. It follows that B and C are symmetric matrices with nonnegative integral elements of orders m and n, respectively. These matrices already reveal a good deal about the internal structure of the subsets. We note that the (i,j) position of B records the cardinality of $X_i \cap X_j$ and the (i,j) position of C records the number of times that the elements x_i and x_j occur among the subsets $X_1, \ldots, X_m$.

One of the most remarkable general theorems on (0,1)-matrices is the following theorem of König. (The theorem is also often referred to as the König-Egerváry theorem and the Frobenius-König theorem.) An enormous literature centers around this theorem and the related theorem of P. Hall [18] on systems of distinct representatives, the theorem of Dilworth [10] on partially ordered sets, and the theorem of Ford and Fulkerson [12] on maximal flows and minimal cuts. A very detailed discussion of these topics appears in the recent book by Mirsky [29]. A short and self-contained proof of König's theorem is available in [33]. Throughout our discussion a *line* of a matrix designates either a row or a column of the matrix.

THEOREM 2.1: *Let A be a $(0,1)$-matrix of size m by n. The minimal number of lines in A that cover all of the 1's in A is equal to the maximal number of 1's in A with no two of the 1's on a line.*

König's theorem has the following set theoretic interpretation. The smallest number of sets and elements that must be deleted from our configuration so that only empty sets remain is the same as the largest number of distinct elements that may be selected from the subsets with the requirement that at most one element is selected from each of the subsets.

3. THE PERMANENT

We now let $A = [a_{ij}]$ denote a matrix of size m by n with elements in a field F and we assume that $m \leq n$. A *diagonal product* of A is the product of m elements of A such that no two of the elements are on a line. The *permanent* of A is defined as the sum of all of the diagonal products of A. Thus we may write

$$\text{per}(A) = \Sigma \, a_{1i_1} a_{2i_2} \cdots a_{mi_m}, \tag{3.1}$$

where the summation in (3.1) extends over all m-permutations $(i_1, i_2, \ldots, i_m)$ of the integers $1, 2, \ldots, n$.

This scalar function of the matrix A appears repeatedly throughout the combinatorial literature. One of the reasons for this is that in case A is a $(0,1)$-matrix then $\text{per}(A)$ counts the number of nonzero diagonal products of A. A detailed survey of permanents is available in [27].

We now deal with the case of square matrices of order n. Then the permanent is the same type of expression as the determinant apart from a factor ± 1 preceding each product on the right side of (3.1). But $\text{per}(A)$ is not multiplicative and the addition of a multiple of one row of A to another does not leave per(A) invariant. These facts make the evaluation of per(A) much more difficult. The following formula for per(A) is a consequence of the principle of inclusion and exclusion [31]. A *row sum* (*line sum*) of a matrix is the sum of the elements in a row (line) of the matrix.

THEOREM 3.1: *Let A be a matrix of order n with elements in a field F. Let A_r denote a matrix obtained from A by replacing r columns of A by zeros. Let $S(A_r)$ denote the product of the row sums of A_r and let $\Sigma S(A_r)$ denote the sums of the $S(A_r)$ over all of the choices for A_r. Then*

$$\text{per}(A) = S(A) - \Sigma S(A_1) + \Sigma S(A_2) - \cdots + (-1)^{n-1}\Sigma S(A_{n-1}). \tag{3.2}$$

Jurkat and Ryser [21] have introduced a very different technique for the evaluation of the permanent of a matrix A of order n. Here per(A) is regarded as a matrix of order 1 and is expressed as a product of n rectangular matrices. Each of the rectangular matrices is formed from a single row of A and has an interesting recursive structure. An analogous matrix factorization holds for det(A). We illustrate these matrix factorization equations for orders 2 and 3:

$$[a_1 \ a_2] \begin{bmatrix} b_2 \\ \pm b_1 \end{bmatrix} = a_1 b_2 \pm a_2 b_1 = \begin{cases} \text{per}(A) \\ \det(A) \end{cases}, A = \begin{bmatrix} a_1 & a_2 \\ b_1 & b_2 \end{bmatrix},$$

$$[a_1 \ a_2 \ a_3] \begin{bmatrix} b_2 & b_3 & 0 \\ \pm b_1 & 0 & b_3 \\ 0 & \pm b_1 & \pm b_2 \end{bmatrix} \begin{bmatrix} c_3 \\ \pm c_2 \\ c_1 \end{bmatrix} = \begin{cases} \text{per}(A) \\ \det(A) \end{cases},$$

$$A = \begin{bmatrix} a_1 & a_2 & a_3 \\ b_1 & b_2 & b_3 \\ c_1 & c_2 & c_3 \end{bmatrix}.$$

We remark that a comparison of the computational effectiveness of the two formulas for per(A) is a topic of some interest in its own right [23].

The matrix factorization equations for per(A) and det(A) suggest the possibility of the existence of a wide variety of matrix identities of great combinatorial interest. In this regard we mention a most remarkable identity of Amitsur and Levitzki [1]. Let A_1, ..., A_k be matrices of order n with elements in a field F. We define

$$[A_1, \ldots, A_k] = \Sigma \text{sgn}(\sigma) A_{\sigma 1} \cdots A_{\sigma k}, \qquad (3.3)$$

where the summation is taken over all permutations σ of the integers $1, \ldots, k$.

THEOREM 3.2: *Let $A_1, \ldots, A_{2n}$ be matrices of order n with elements in a field F. Then*

$$[A_1, \ldots, A_{2n}] = 0. \qquad (3.4)$$

A combinatorial proof of the Amitsur-Levitzki identity based on graph theory has been given by Swan [40], [41]. The Amitsur-Levitzki identity is of "determinant" type. It would be of the utmost interest to have related matrix identities of "permanent" type.

The most celebrated topic in the theory of permanents is known as the van der Waerden conjecture. A matrix A of order n is called *doubly stochastic* provided that the elements of A are nonnegative real numbers and all of the line sums of A are equal to 1. The van der Waerden conjecture asserts that every doubly stochastic matrix A of order n satisfies

$$\text{per}(A) \geq \frac{n!}{n^n}. \qquad (3.5)$$

Equality holds in (3.5) for $A = n^{-1}J$, where J is the matrix of 1's of order n. This may be in fact the only case of equality in (3.5).

The validity of the van der Waerden conjecture has been verified for only the first few values of n.

The following elegant theorem of Marcus and Newman [28] adds some support to the possible validity of the van der Waerden conjecture:

THEOREM 3.3: *Let A be a symmetric positive semidefinite doubly stochastic matrix of order n. Then A satisfies*

$$\mathrm{per}(A) \geq \frac{n!}{n^n}, \qquad (3.6)$$

and equality holds in (3.6) if and only if $A = n^{-1}J$.

A number of interesting inequalities and conjectures also arise in the study of permanents of (0,1)-matrices. Let A be a (0,1)-matrix of order n such that all of the line sums of A are equal to k, where k is a fixed integer in the interval $1 \leq k \leq n$. Then an inequality of M. Hall [16] asserts that

$$\mathrm{per}(A) \geq k! . \qquad (3.7)$$

But the actual minimal value of per(A) as a function of k is far from understood. Indeed, we may regard the determination of this minimal value of per(A) as the "discrete version" of van der Waerden's original problem.

4. SYMMETRIC BLOCK DESIGNS

We now define configurations that play a very central role in combinatorial matrix theory. The subsets $X_1, \ldots, X_v$ of a v-set $X = \{x_1, \ldots, x_v\}$ are called a (v,k,λ)-*design* (*symmetric block design*) provided that they satisfy the following postulates:

$$\text{Each } X_i \text{ is a } k\text{-subset of } X. \qquad (4.1)$$

$$\text{Each } X_i \cap X_j \text{ for } i \neq j \text{ is a } \lambda\text{-subset of } X. \qquad (4.2)$$

COMBINATORIAL MATRIX THEORY

The integers v, k, and λ satisfy $0 < \lambda < k < v - 1$. (4.3)

The above postulates imply that the incidence matrix A of a (v, k, λ)-design is a $(0,1)$-matrix of order v that satisfies the matrix equation

$$AA^T = (k - \lambda)I + \lambda J. \quad (4.4)$$

In (4.4) A^T is the transpose of the matrix A, I is the identity matrix of order v, and J is the matrix of 1's of order v.

We now prove that the incidence matrix A of a (v, k, λ)-design is normal, namely,

$$AA^T = A^T A. \quad (4.5)$$

An elementary calculation tells us that

$$\det((k - \lambda)I + \lambda J) = (k - \lambda + \lambda v)(k - \lambda)^{v-1} \neq 0. \quad (4.6)$$

Hence it follows that A is a nonsingular matrix and we denote the inverse of A by A^{-1}. Thus we may write

$$AJ = kJ, A^{-1}J = k^{-1}J. \quad (4.7)$$

We multiply (4.4) on the right by J and this gives

$$AA^T J = (k - \lambda + \lambda v)J. \quad (4.8)$$

Then (4.7) and (4.8) imply

$$A^T J = (k - \lambda + \lambda v)k^{-1}J. \quad (4.9)$$

We next take the transpose of both sides of (4.9) and obtain

$$JA = (k - \lambda + \lambda v)k^{-1}J. \quad (4.10)$$

Hence we have

$$JAJ = (k - \lambda + \lambda v)k^{-1}vJ. \quad (4.11)$$

But by (4.7) we also have

$$JAJ = kvJ, \tag{4.12}$$

whence we obtain the interesting relationship

$$k - \lambda = k^2 - \lambda v. \tag{4.13}$$

Then (4.10) and (4.13) imply

$$JA = kJ. \tag{4.14}$$

Finally, we write

$$\begin{aligned} A^T A &= A^{-1}(AA^T)A \\ &= (k - \lambda)I + \lambda A^{-1}JA \\ &= (k - \lambda)I + \lambda J, \end{aligned} \tag{4.15}$$

and this is the desired conclusion.

One of the major unsolved problems in combinatorics is the determination of the precise range of values of v, k, and λ for which (v, k, λ)-designs exist. It is already clear that the parameters v, k, and λ are far from arbitrary because they must satisfy not only (4.3) but also (4.13). The only other known necessary conditions on the parameters are those given by the following Bruck-Ryser-Chowla theorem on the existence of (v, k, λ)-designs:

THEOREM 4.1: *Suppose that v, k, and λ are integers for which there exists a (v, k, λ)-design. If v is even then the integer $k - \lambda$ is equal to a square. If v is odd then the Diophantine equation*

$$x^2 = (k - \lambda)y^2 + (-1)^{(v-1)/2}\lambda z^2 \tag{4.16}$$

has a solution in integers x, y, and z not all equal to zero.

The case of v even follows at once from the fact that the determinant evaluated in (4.6) must be equal to a square. But the case of v odd requires a deeper analysis of matrix congruences over the field of rational numbers.

Certain special parameter sets on v, k, and λ are of great importance in their own right. A *finite projective plane* of *order n* is a (v, k, λ)-design on the parameters

$$v = n^2 + n + 1, k = n + 1, \lambda = 1 \quad (n \geq 2). \tag{4.17}$$

Finite projective planes are the counterparts in geometry of the finite fields in algebra. But all finite fields are known explicitly, whereas the structure of finite planes is far from understood.

Finite planes are easily constructed from finite fields in case the order of the plane is equal to the power of a prime number. It is conjectured that the order of a finite plane must be equal to the power of a prime number. The first undecided order is $n = 10$. The construction of a plane of order 10 is equivalent to finding a (0,1)-matrix A of order 111 such that AA^T has 11 in all of the main diagonal positions and 1 in all other positions. This problem has been under intensive investigation [26] and is perhaps the most famous unsolved problem that is of a purely finite character.

We now mention certain matrices that lead to another important class of (v, k, λ)-designs. Let H be a matrix of order n such that the elements of H are the integers 1 and -1. Suppose that H satisfies the matrix equation

$$HH^T = nI, \tag{4.18}$$

where H^T is the transpose of H and I is the identity matrix of order n. Then H is called a *Hadamard matrix* of order n. It is elementary to verify that a Hadamard matrix must have order $n = 1$, 2, or else $n \equiv 0 \pmod 4$. It is conjectured that Hadamard matrices exist for all orders $n \equiv 0 \pmod 4$. They have been constructed for numerous values of n.

A row or a column of a Hadamard matrix may be multiplied by -1 and the resulting matrix is again a Hadamard matrix. Hence we may assume that the Hadamard matrix has its first row and column composed entirely of 1's. Such a Hadamard matrix is called *normalized*. Suppose now that H is a normalized Hadamard matrix of order $n = 4t \geq 8$. Then we may delete row 1 and column 1 of H and replace the -1's in H by 0's. We thereby obtain a $(0, 1)$-matrix A of order $v = 4t - 1$. It follows directly from the structure of A that a normalized Hadamard matrix of order $n = 4t \geq 8$ is equivalent to a (v, k, λ)-design on the parameters

$$v = 4t - 1, k = 2t - 1, \lambda = t - 1 \qquad (t \geq 2). \tag{4.19}$$

Such a design is called a *Hadamard design*. Thus the main conjecture on Hadamard matrices is equivalent to the conjecture that Hadamard designs exist for all choices of the parameters (4.19).

Another special class of (v, k, λ)-designs that has received considerable attention of late are those with the parameter $\lambda = 2$. Such designs are called *biplanes* and they are exceedingly rare [8]. Only a few such configurations are known. It is in fact conjectured that the number of (v, k, λ)-designs for each fixed value of $\lambda > 1$ is finite.

We make no attempt to summarize further the enormous literature on symmetric block designs. Such summaries are available in the books by Dembowski [9], Hall [17], and Ryser [31].

5. RECENT VARIANTS OF SYMMETRIC BLOCK DESIGNS

In the definition of a (v, k, λ)-design postulate (4.3) is unimportant because it merely excludes certain degenerate configurations, whereas postulate (4.2) is entirely basic to the definition of these configurations. The status of postulate (4.1) is somewhat less clear. In the following theorem of Ryser [32] and Woodall [43] the scalar matrix on the right side of (4.4) is replaced by a diagonal matrix and we obtain information in the event that postulate (4.1) is denied.

THEOREM 5.1: *Let A be a $(0, 1)$-matrix of order $n > 1$ that satisfies the matrix equation*

$$AA^T = \text{diag}[k_1 - \lambda, \ldots, k_n - \lambda] + \lambda J, \tag{5.1}$$

where A^T is the transpose of A and J is the matrix of 1's of order n. Suppose that not all of the k_i are equal and that $0 < \lambda < k_i$. Then A has exactly two distinct column sums c_1 and c_2 and these numbers satisfy

$$c_1 + c_2 = n + 1. \tag{5.2}$$

The configurations associated with the incidence matrix A of Theorem 5.1 are variants of symmetric block designs and they are called λ-*designs* on n elements. The λ-designs with $\lambda = 1$ have an especially simple structure. One may prove that for each $n > 3$ there exists a unique λ-design with $\lambda = 1$ [7]:

$$X_1 = \{2, 3, \ldots, n\},$$
$$X_2 = \{1, 2\}, X_3 = \{1, 3\}, \ldots, X_n = \{1, n\}.$$

This is in sharp contrast to the state of affairs for a finite projective plane. One may also prove that there exists a unique λ-design with $\lambda = 2$ [32]:

$$\{1, 2, 4\}, \{1, 4, 6, 7\}, \{2, 5, 7, 1\}, \{3, 6, 1, 2\}, \{4, 7, 2, 3\},$$
$$\{5, 1, 3, 4\}, \{6, 2, 4, 5\}.$$

A simple but judicious modification of the incidence matrix of a symmetric block design allows us to construct a λ-design [2], [43]. All of the λ-designs constructed by this procedure (including the λ-designs with $\lambda = 1$) are called λ-designs of *type* 1. It is conjectured that all λ-designs are of type 1 and the validity of this conjecture has been verified for all $\lambda \leq 9$ [2], [4], [25].

Thus we see that the role of postulate (4.1) in the definition of a symmetric block design is not fully understood. Its deletion leaves open the possibility of the existence of exotic new configurations.

The following theorem of Woodall [43] settles for λ-designs a question that we raised earlier for (v, k, λ)-designs:

THEOREM 5.2: *The number of λ-designs for each fixed value of $\lambda > 1$ is finite.*

A λ-design may be regarded as a variant of a symmetric block design obtained by a modification of the right side of the matrix equation (4.4). The following theorem of Bridges and Ryser [5] is concerned with a modification of the left side of the matrix equation (4.4):

THEOREM 5.3: *Let X and Y be nonnegative integral matrices of order $n > 1$ such that*

$$XY = (k - \lambda)I + \lambda J, \tag{5.3}$$

where I is the identity matrix of order n and J is the matrix of 1's of order n. Suppose that $k \ne \lambda$ and that the integers k and λ are relatively prime. Then there exist positive integers r and s such that X has all of its line sums equal to r and Y has all of its line sums equal to s, where

$$rs = k + (n - 1)\lambda. \tag{5.4}$$

Moreover,

$$XY = YX. \tag{5.5}$$

We remark that the matrix equation (5.3) has a set theoretic interpretation in its own right and that certain special solutions of (5.3) lead to configurations that are known as strongly regular graphs [13], [39].

We note that the assumption $k \ne \lambda$ in Theorem 5.3 is an essential part of the hypothesis. Suppose, for example, that we take a square R of side n and that we partition R into n smaller rectangles, each of which has integral height and integral length. Then by Section 1 this partition of R yields two (0,1)-matrices X and Y of order n such that

$$XY = J. \tag{5.6}$$

But neither X nor Y need have all of its line sums equal.

We now let A denote a $(0,1)$-matrix of order n and we suppose that A satisfies the matrix equation

$$A^2 = J. \tag{5.7}$$

This equation arises naturally in the study of central groupoids, universal algebras, and graphs with exactly one path of length two between every pair of points [24]. We do not pursue these applications here. But it is important to bear in mind that the various matrix equations under discussion have far reaching implications.

We now inquire into the structure of a $(0,1)$-matrix A of order n that satisfies the matrix equation (5.7). We multiply (5.7) on the left and the right by A and this gives

$$JA = AJ = A^3. \tag{5.8}$$

Then multiplication of (5.8) on the left by J implies

$$J^2 A = nJA = JAJ = eJ, \tag{5.9}$$

where e denotes the sum of all of the elements of A. Thus A has all of its line sums equal to

$$c = \frac{e}{n}. \tag{5.10}$$

But then multiplication of (5.7) on the left (or the right) by J implies

$$c^2 = n. \tag{5.11}$$

Hence it follows that a $(0,1)$-matrix A of order n that satisfies (5.7) must have n equal to a square.

It is easy to construct a "natural" solution of (5.7) whenever n is equal to a square. We illustrate the natural solution for $n = 4$:

$$\begin{bmatrix} 1 & 1 & 0 & 0 \\ 0 & 0 & 1 & 1 \\ 1 & 1 & 0 & 0 \\ 0 & 0 & 1 & 1 \end{bmatrix}.$$

At this point one is tempted to classify all of the solutions of (5.7). This problem has been proposed by Hoffman [19] but remains unsolved.

We carry out one further calculation concerning the (0,1)-matrix A of (5.7). The matrix J has a characteristic root n of multiplicity 1 and a characteristic root 0 of multiplicity $n - 1$. Hence by wellknown properties of characteristic roots it follows that A has a characteristic root $\pm c$ of multiplicity 1 and a characteristic root 0 of multiplicity $n - 1$. But the sum of the n characteristic roots of A is the sum of the main diagonal of A so that c and not $-c$ is a characteristic root of A. We conclude that an arbitrary (0,1)-matrix A of order n that satisfies (5.7) must have exactly c 1's on its main diagonal.

Various extensions and modifications of the matrix equation (5.7) have been studied and we cite the following references: [3], [20], [34].

6. INDETERMINATES AND INCIDENCE MATRICES

We return to the subsets $X_1, \ldots, X_m$ of an n-set $X = \{x_1, \ldots, x_n\}$ described at the outset of Section 2. We recall that the incidence matrix for these subsets of X is the (0,1)-matrix A of size m by n with the property that row i of A displays the subset X_i and column j of A displays the occurrences of the element x_j among the subsets.

We now regard the elements of the n-set $X = \{x_1, \ldots, x_n\}$ as independent indeterminates over the field of rational numbers and we also denote by X the diagonal matrix of order n

$$X = \text{diag}[x_1, \ldots, x_n]. \tag{6.1}$$

In a recent paper Ryser [35] introduced the matrix equation

$$AXA^T = Y, \tag{6.2}$$

where A^T denotes the transpose of the matrix A.

The matrix equation (6.2) contains a wealth of information in an exceedingly compact form. The matrix Y is a symmetric matrix of order m and we know the structure of this matrix explicitly. Thus the matrix Y has in its (i, j) position the sum of the indeterminates in $X_i \cap X_j$. It follows that the matrix Y gives us an explicit representation for all of the set intersections $X_i \cap X_j$ of the subsets $X_1, \ldots, X_m$ of the n-set X.

The matrix equation (6.2) is useful in dealing with certain types of set intersection problems [35] and the algebraic properties of a somewhat more general matrix equation have also been studied [37]. Earlier investigations on matrices and set intersections include [14], [15], [22], [36].

We note a few elementary properties of the matrix equation (6.2). The standard theorems on the rank of a matrix imply

$$\text{rank}(Y) = \text{rank}(A), \tag{6.3}$$

and in the special case $m = n$ we may take determinants in (6.2) and this gives

$$\det(Y) = (\det(A))^2 \, \Pi^n_{i=1} x_i. \tag{6.4}$$

The matrix equation (6.2) involves indeterminates and we may assign these indeterminates arbitrary rational values. Each such assignment gives us a matrix equation that must be satisfied by the incidence matrix A for our configuration of subsets. Thus if we set $x_1 = \cdots = x_n = 1$ then (6.2) reduces to our earlier equation (2.3), and the matrix B of (2.3) reveals the cardinalities of the set intersections. Highly significant applications of the matrix equation (6.2) are not out of the question. These could involve an ingenious assignment of values to the indeterminates so that new insights into the unsolved problems of Sections 4 and 5 are revealed.

We now discuss a rather different usage for indeterminates in combinatorial matrix theory. König's theorem described in Section 2 deals with a (0,1)-matrix of size m by n. The theorem has an immediate extension to an arbitrary matrix A of size m by n with elements in a field F.

THEOREM 6.1: *Let A be a matrix of size m by n with elements in a field F. The minimal number of lines in A that cover all of the nonzero elements in A is equal to the maximal number of nonzero elements in A with no two of the nonzero elements on a line.*

The maximal number of nonzero elements in A with no two of the nonzero elements on a line is called the *term rank* of A. We now disregard our earlier notation and let

$$X = [x_{ij}] \quad (i = 1, \ldots, m; \quad j = 1, \ldots, n) \qquad (6.5)$$

denote the matrix of size m by n, where the elements of X are mn independent indeterminates over F. We call the Hadamard product

$$M = A * X = [a_{ij} x_{ij}] \qquad (6.6)$$

the *formal incidence matrix* associated with A. The elements of M belong to the polynominal ring

$$F^* = F[x_{11}, x_{12}, \ldots, x_{mn}]. \qquad (6.7)$$

The formal incidence matrix turns out to be very useful in dealing with various combinatorial investigations [6], [11], [30], [38], [42]. One of the reasons for this is that an important combinatorial invariant of A is equal to an algebraic invariant of M. *The term rank of A is equal to the rank of M.* This observation of Edmonds [11] may be derived as follows. The definition of the formal incidence matrix M implies that a submatrix of M of order r has a nonzero determinant if and only if the corresponding submatrix of A has term rank r. But the rank of a matrix is equal to the maximal order of a square submatrix with a nonzero determinant. Hence the conclusion follows. Thus, for example, a (0,1)-matrix A of size m by n with $m \leq n$ has per$(A) > 0$ if and only if M is of rank m.

We now let A denote a matrix of order $n > 1$ with elements in a field F. The matrix A is *fully indecomposable* provided that it does not contain a zero submatrix of size r by $n - r$, for some integer r such that $1 \leq r \leq n - 1$. Thus the nonzero elements of a fully indecomposable matrix A of order $n > 1$ cannot be covered by n lines that are composed of both rows and columns of A. If the matrix A is of order $n = 1$ then we say that A is *fully indecomposable* provided that A is not the zero matrix of order 1. An immediate consequence of Theorem 6.1 is the assertion that a fully indecomposable matrix A of order n has term rank n and hence det$(M) \neq 0$. But det$(M) \neq 0$ does not in general imply that the matrix A is fully indecomposable. However, the following theorem already evident from the investigations of Frobenius shows that a fully indecomposable matrix A is characterized by an algebraic property of det(M) [38].

THEOREM 6.2: *Let A be a matrix of order n with elements in a field F and let $M = A * X$ be the formal incidence matrix associated with A. Then A is fully indecomposable if and only if* det(M) *is an irreducible polynomial in the polynomial ring*

$$F^* = F[x_{11}, x_{12}, \ldots, x_{nn}]. \tag{6.8}$$

In this section we have indicated two uses for indeterminates in combinatorial matrix theory. However, we anticipate that many other uses will emerge and that matrices in conjunction with indeterminates will provide an exceedingly effective device for dealing with varied combinatorial problems.

REFERENCES

1. S. A. Amitsur and J. Levitzki, "Minimal identities for algebras", *Proc. Amer. Math. Soc.*, **1** (1950), 449-463.
2. W. G. Bridges, "Some results on λ-designs", *J. Combinatorial Theory*, **8** (1970), 350-360.

3. ____, "The polynomial of a non-regular digraph", *Pacific J. Math.*, **38** (1971), 325-341.
4. W. G. Bridges and E. S. Kramer, "The determination of all λ-designs with λ = 3", *J. Combinatorial Theory*, **8** (1970), 343-349.
5. W. G. Bridges and H. J. Ryser, "Combinatorial designs and related systems", *J. Algebra*, **13** (1969), 432-446.
6. R. A. Brualdi and H. Perfect, "Extension of partial diagonals of matrices I", *Monatsh. Math.*, **75** (1971), 385-397.
7. N. G. de Bruijn and P. Erdös, "On a combinatorial problem", *Indag. Math.*, **10** (1948), 421-423.
8. P. J. Cameron, "Biplanes", *Math. Z.*, **131** (1973), 85-101.
9. P. Dembowski, *Finite Geometries*, Springer-Verlag, Berlin, 1968.
10. R. P. Dilworth, "A decomposition theorem for partially ordered sets", *Ann. of Math.*, (2) **51** (1950), 161-166.
11. J. Edmonds, "Systems of distinct representatives and linear algebra", *J. Res. Nat. Bur. Standards*, **71B** (1967), 241-245.
12. L. R. Ford, Jr., and D. R. Fulkerson, *Flows in Networks*, Princeton University Press, 1962.
13. J. M. Goethals and J. J. Seidel, "Orthogonal matrices with zero diagonal", *Canad. J. Math.*, **19** (1967), 1001-1010.
14. A. W. Goodman, "Set equations", *Amer. Math. Monthly*, **72** (1965), 607-613.
15. M. Hall, Jr., "A problem in partitions", *Bull. Amer. Math. Soc.*, **47** (1941), 804-807.
16. ____, "Distinct representatives of subsets", *Bull. Amer. Math. Soc.*, **54** (1948), 922-926.
17. ____, *Combinatorial Theory*, Blaisdell, Waltham, Mass., 1967.
18. P. Hall, "On representatives of subsets", *J. London Math. Soc.*, **10** (1935), 26-30.
19. A. J. Hoffman, "Research problems", *J. Combinatorial Theory*, **2** (1967), 393.
20. A. J. Hoffman and M. H. McAndrew, "The polynomial of a directed graph", *Proc. Amer. Math. Soc.*, **16** (1965), 303-309.
21. W. B. Jurkat and H. J. Ryser, "Matrix factorizations of determinants and permanents", *J. Algebra*, **3** (1966), 1-27.
22. J. B. Kelly, "Products of zero-one matrices", *Canad. J. Math.*, **20** (1968), 298-329.
23. D. E. Knuth, *The Art of Computer Programming*, Vol. 2, Seminumerical Algorithms, Addison-Wesley, Reading, Mass., 1969.
24. ____, "Notes on central groupoids", *J. Combinatorial Theory*, **8** (1970), 376-390.
25. E. S. Kramer, "On λ-designs", Dissertation, University of Michigan, 1969.
26. F. J. MacWilliams, N. J. A. Sloane, and J. G. Thompson, "On the existence of a projective plane of order 10", *J. Combinatorial Theory*, **A14** (1973), 66-78.
27. M. Marcus and H. Minc, "Permanents", *Amer. Math. Monthly*, **72** (1965), 577-591.
28. M. Marcus and M. Newman, "Inequalities for the permanent function", *Ann. of Math.*, **75** (1962), 47-62.
29. L. Mirsky, *Transversal Theory*, Academic Press, New York, 1971.

30. H. Perfect, "Symmetrized form of P. Hall's theorem on distinct representatives", *Quart. J. Math. Oxford Ser.*, (2) **17** (1966), 303-306.
31. H. J. Ryser, *Combinatorial Mathematics*, Carus Math. Monograph No. 14, Mathematical Association of America, 1963.
32. ____, "An extension of a theorem of de Bruijn and Erdös on combinatorial designs", *J. Algebra*, **10** (1968), 246-261.
33. ____, "Combinatorial configurations", *SIAM J. Appl. Math.*, **17** (1969), 593-602.
34. ____, "A generalization of the matrix equation $A^2 = J$", *Linear Algebra and Appl.*, **3** (1970), 451-460.
35. ____, "A fundamental matrix equation for finite sets", *Proc. Amer. Math. Soc.*, **34** (1972), 332-336.
36. ____, "Intersection properties of finite sets", *J. Combinatorial Theory*, **A14** (1973), 79-92.
37. ____, "Analogs of a theorem of Schur on matrix transformations", *J. Algebra*, **25** (1973), 176-184.
38. ____, "Indeterminates and incidence matrices", *Linear and Multilinear Algebra*, **1** (1973), 149-157.
39. J. J. Seidel, "Strongly regular graphs with $(-1, 1, 0)$ adjacency matrix having eigenvalue 3", *Linear Algebra and Appl.*, **1** (1968), 281-298.
40. R. G. Swan, "An application of graph theory to algebra", *Proc. Amer. Math. Soc.*, **14** (1963), 367-373.
41. ____, "Correction to 'An application of graph theory to algebra'", *Proc. Amer. Math. Soc.*, **21** (1969), 379-380.
42. W. T. Tutte, "The factorization of linear graphs", *J. London Math. Soc.*, **22** (1947), 107-111.
43. D. R. Woodall, "Square λ-linked designs", *Proc. London Math. Soc.*, (3) **20** (1970), 669-687.

PROOF TECHNIQUES IN THE THEORY OF FINITE SETS*

Curtis Greene and Daniel J. Kleitman

1. INTRODUCTION

This paper is a survey of a class of methods which have proved successful in attacking certain problems in the theory of finite sets. Specifically, we will be concerned with *extremal problems*: i.e., given a family F of finite sets, satisfying certain restrictions, how large can F be? Typically, an answer to this question leads to a classification of the extremal cases, so that many of the results in this paper are ultimately structural as well as numerical.

No attempt has been made to provide a complete survey of the field: many beautiful results which do not fit into our (perhaps rather arbitrary) scheme have been omitted entirely. And although we have included a wide variety of problems, our motivation in choosing them has been at least partially based on how well they illustrate certain techniques. For a more encyclopedic treatment, we recommend the survey articles by Katona [30] and Erdös and Kleitman [16].

---

*Supported in part by ONR N00014-67-A-0204-0063.

Because our goal is to study techniques, we shall also investigate the extent to which these techniques can be generalized to a wider class of combinatorial objects. Specifically, we will attempt—wherever possible—to extend results about families of sets to more general kinds of partially ordered structures, with a special emphasis on three important combinatorial examples: *multisets* (or *divisors of an integer*), *subspaces of a finite vector space*, and *partitions of a set*. The structures formed by these combinatorial objects are analogous in many ways to Boolean algebras of sets, and provide a rich source of problems. By using them as examples, we hope to illustrate both the power and the limitations of the methods described.

It should be mentioned that our systematic treatment of the analogies between families of finite sets and other combinatorial objects is very much in the spirit of ideas originally suggested by Rota (see, for example, [21], [51], or [63]).

THE SPERNER PROBLEM

If P is a partially ordered set, an *antichain of P* is a subset of P whose elements are totally unrelated (as opposed to a *chain*, whose elements are totally related). One of the earliest results of the kind considered in this paper is a theorem about antichains of sets, due to E. Sperner and published in 1928:

THEOREM 1.1 (Sperner): *Let F be a family of subsets of $\{1, 2, \ldots, n\}$, no member of which contains another. Then $|F| \leq \binom{n}{\left[\frac{n}{2}\right]}$. Equality occurs only if all of the sets have the same size.*

Given an arbitrary partially ordered set P, we can ask a similar question: what is the size of the largest antichain in P? In this paper, we will describe a number of ways to approach this problem, and many of the results discussed can be viewed as extensions, refinements, or analogs of Sperner's fundamental theorem.

For arbitrary partially ordered sets, the problem of finding a maximum-sized antichain can be expressed as a network flow problem and solved (efficiently) by standard techniques (see [58] for a description of an elementary algorithm). We will not be concerned with the problem in such generality, but rather with special cases where the answer has a simple form.

Suppose that P is a partially ordered set with a *rank function* r. That is, r is a function defined on the elements of P, taking nonnegative integer values, such that $r(x) = 0$ for every minimal element x, and $r(x_1) = r(x_2) + 1$ whenever x_1 covers x_2. For each integer k, let P_k denote the collection of elements in P having rank k. Clearly each P_k is an antichain. When P consists of all subsets of a finite set, rank coincides with cardinality, and Sperner's theorem asserts than an antichain of maximum size can be obtained by taking all elements of rank $\left[\dfrac{n}{2}\right]$. In general, we say that P has the *Sperner Property* if the maximum size of an antichain in P is equal to $\max_k |P_k|$.

WHITNEY NUMBERS

The following notation will be useful when considering more general classes of partially ordered sets: if P has a rank function, let $N_k(P)$ denote the number of elements of rank k in P (i.e., $N_k(P) = |P_k|$). Following Crapo and Rota [8], we call $N_k(P)$ the *kth Whitney number of P (of the second kind)*.[*]

There are four basic families of partially ordered sets considered in this paper: *sets, multisets, subspaces,* and *partitions*. We describe each of them briefly below, including a calculation of the appropriate Whitney numbers for each class:

(1) *Sets*: Let B_n denote the Boolean algebra of all subsets of $\{1, 2, \ldots, n\}$, ordered by inclusion. Then, as noted before, B has a rank function $r(s) = |S|$, and

---

[*] When there is no chance of confusion, we will write $N_k(P) = N_k$, and if $x \in P$ we also write $N_{r(x)}(P) = N_x(P) = N_x$.

$$N_k(B_n) = \frac{n!}{k!\,(n-k)!} = \binom{n}{k}$$

(2) *Multisets*: Fix a sequence $\vec{e} = (e_1, e_2, \ldots)$, where each e_i is either a nonnegative integer or $+\infty$. Let $M_{\vec{e}}$ denote the collection of all "finite multisets of integers with multiplicities restricted by $\vec{e}$." By this we mean the family of all finite unordered collections of positive integers, where repetitions are allowed but each i can appear at most e_i times. Each multiset in $M_{\vec{e}}$ can be represented by a sequence $\vec{\sigma} = (\sigma_1, \sigma_2, \ldots,)$ of nonnegative integers σ_i such that $0 \le \sigma_i \le e_i$ for each i and also $\Sigma \sigma_i < \infty$. For two multisets $\vec{\sigma}$ and $\vec{\sigma}'$, define $\vec{\sigma} \le \vec{\sigma}'$ if $\sigma_i \le \sigma_i'$ for all i. Under this ordering $M_{\vec{e}}$ is a distributive lattice—in fact $M_{\vec{e}}$ is isomorphic to a cartesian product of chains.

It is easy to see that $M_{\vec{e}}$ has a rank function given by $r(\vec{\sigma}) = \Sigma \sigma_i$, and that $N_k(M_{\vec{e}})$ is the coefficient of x^k in the expression

$$\prod_{i=1}^{\infty}(1 + x + \cdots + x^{e_i})$$

(with the obvious convention if $e_i = \infty$). We also introduce the notation

$$N_k(M_{\vec{e}}) = \binom{e_1, e_2, \ldots}{k}.$$

Although these numbers are more difficult to work with than binomial coefficients, one can obtain a trivial analog of the binomial recursion: if $\vec{e}\,'$ is obtained from $\vec{e}$ by replacing e_m by 0, for some m, then

$$\binom{e_1, e_2, \ldots}{k} = \sum_{i=0}^{e_m} \binom{e_1', e_2', \ldots}{k - i}$$

(where, by convention, a term vanishes if $k - i$ is negative).

When $e_1 = e_2 = \cdots = e_n = \infty$, $e_{n+1} = e_{n+2} = \cdots = 0$, we call $M_{\vec{e}}$ the *lattice of unrestricted (finite) multisubsets of* $\{1, 2, \ldots, n\}$. In this case, $N_k(M_{\vec{e}})$ is the coefficient x^k in the expression $(1 - x)^{-n}$, and we obtain the explicit formula

$$N_k(M_{\vec{e}}) = (-1)^k \binom{-n}{k} = \binom{n + k - 1}{k}.$$

(2') *Divisors of an integer*: If N is any positive integer, let D_N denote the lattice of divisors of N, ordered by the relation of divisibility. If $N = p_1^{e_1} p_2^{e_2} \cdots p_k^{e_k}$ is the prime decomposition of N, the divisors of N can be thought of as multisets of primes, with multiplicities restricted by $e_1, e_2, \ldots, e_k$, and the ordering of multisets coincides with that of divisors. Hence D_N is isomorphic to $M_{\vec{e}}$, where $\vec{e} = (e_1, e_2, \ldots, e_k, 0, 0, \ldots)$.

If we take $e_i = \infty$ for all i, then $M_{\vec{e}}$ is isomorphic to the lattice of all positive integers, ordered by divisibility.

(3) *Subspaces*: Let $L_n(q)$ denote the lattice of subspaces of a vector space of dimension n over a field of q elements, ordered by inclusion. Then $L_n(q)$ has a rank function $r(U) = \dim(U)$, and

$$N_k(L_n(q)) = \frac{(q^n - 1)(q^{n-1} - 1) \cdots (q^{n-k+1} - 1)}{(q^k - 1)(q^{k-1} - 1) \cdots (q - 1)} = \begin{bmatrix} n \\ k \end{bmatrix}_q.$$

These numbers are called *Gaussian coefficients*, and are polynomials in q for fixed n and k, as can be seen from the recursion

$$\begin{bmatrix} n + 1 \\ k \end{bmatrix}_q = \begin{bmatrix} n \\ k - 1 \end{bmatrix}_q + q^k \begin{bmatrix} n \\ k \end{bmatrix}_q.$$

It is interesting to note that

$$\lim_{q \to 1} \begin{bmatrix} n \\ k \end{bmatrix}_q = \binom{n}{k}.$$

(4) *Partitions*: Let Π_n denote the collection of all partitions of the set $\{1, 2, \ldots, n\}$, ordered by *refinement*. That is, $\sigma \leq \tau$ if σ can be obtained by subdividing the blocks of τ. It is well known that Π_n is a lattice, with a rank function r defined by $r(\sigma) = n - |\sigma|$, where $|\sigma|$ denotes the number of blocks of σ. The Whitney numbers of Π_n are given by

$$N_k(\Pi_n) = S(n, n - k),$$

where $S(n, n - k)$ denotes a *Stirling number* (of the second kind). Although no explicit formula for the Stirling numbers is known, it is easy to derive the recursion

$$S(n + 1, k) = S(n, k - 1) + k\, S(n, k).$$

Among the four classes of partially ordered sets described here, it is known (and we shall prove) that three have the Sperner Property: sets, multisets, and subspaces. For partitions, however, the answer is still not known, and is the subject of a long-standing conjecture of Rota [51]. (**Note added in proof:** Rota's conjecture has recently been settled in the negative by E. R. Canfield.)

2. SYMMETRY

We begin this section by giving one of the shortest known proofs of Sperner's theorem (although it turns out to be one of the least capable of generalization). The proof rests on the following observation, due to Kleitman, Edelberg, and Lubell [43]:

LEMMA 2.0: *Every partially ordered set P contains an antichain of maximum size which is invariant under every order-automorphism of P.*

To see that Sperner's theorem follows from this fact, observe that any invariant antichain in B_n must consist of all sets of some fixed size i. Such a family has $\binom{n}{i}$ members, which is maximized when $i = [n/2]$.

Freese [20] observed that Lemma 2.0 can be proved in the following way: define an ordering on antichains of P by saying $A \le B$ if every member of A is dominated by some member of B. If A and B are two antichains, define $A \vee B$ to be the antichain of maximal elements in the set $A \cup B$. Trivially $A \le A \vee B$ and $B \le A \vee B$. Moreover if A and B are antichains of maximum size, it is easy to see that $A \vee B$ is again of maximum size. Repeating this operation, we obtain a maximum-sized antichain which is "largest" in the sense of the ordering just defined. This antichain must be invariant under every automorphism of P, and the lemma is proved.

Freese's argument is based on ideas of Dilworth [13], who showed that the maximum-sized antichains of a partially ordered set P form a distributive lattice, which is a sublattice of the (distributive) lattice of all antichains in P.

It is clear that the same arguments prove the following:

THEOREM 2.1: *If P is a partially ordered set with rank function whose automorphism group is transitive on each set of elements of fixed rank, then P has the Sperner property.*

As a consequence, we obtain:

COROLLARY 2.2: *For each n and q, $L_n(q)$ has the Sperner property.*

Unfortunately, the hypotheses of Theorem 2.1 do not hold (in general) for either lattices of multisets or lattices of partitions.

A more general class of problems to which these methods apply can be described as follows: a subset $A \subseteq P$ is called a *k-family* of P if A contains no chains of length $k + 1$. Erdös proved the following [15]:

THEOREM 2.3: *Let F be a family of subsets of $\{1, 2, \ldots, n\}$ that contains no chains of length $k + 1$. Then $|F|$ is bounded by the sum of the k largest binomial coefficients.*

Greene and Kleitman [23] showed that Theorem 2.3 could also be proved by an argument of the Dilworth-Kleitman-Freese type. If A and B are k-families of P, define A_i and B_i to be the elements of "depth" i in A and B respectively (the depth of an element $x \in A$ is the length of the longest chain in A whose bottom is x). Define $A \leq B$ if $A_i \leq B_i$ for $i = 1, 2, \ldots, k$. Then the following can be proved:

THEOREM 2.4: *If P is any partially ordered set, then for each k there exists a k-family of maximum size which is largest with respect to the ordering just defined, and hence is invariant under every automorphism of P.*

COROLLARY 2.5: *The analog of Theorem 2.3 holds in $L_n(q)$ for every n and q, and, more generally, in any partially ordered set whose automorphism group is transitive on elements of a given rank.*

The proof of theorem 2.4 is not immediate for $k > 1$. In section 4 we will see that easier proofs of Erdös' theorem and its analogs are available.

3. SATURATED PARTITIONS

In this section, we describe a method for proving Sperner's Theorem which applies to lattices of sets and lattices of multisets, but (apparently) not to lattices of subspaces or partitions.

Let P be an arbitrary partially ordered set, and let $\mathcal{C} = \{C_1, C_2, \ldots, C_q\}$ be a partition of P into chains C_i. Then $\mathcal{C}$ determines a bound on the size of the largest antichain in P: since chains and antichains have at most one element in common, no antichain in P can have more than q elements. When the bound determined by $\mathcal{C}$ is exact, we call $\mathcal{C}$ a *saturated partition of P*.

By a famous theorem of Dilworth [14], saturated partitions always exist. This means that, if an antichain $A \subseteq P$ is of maximum size, it is always possible to verify this by finding a partition of P into $|A|$ chains.

To prove Sperner's Theorem by this method, we must find a way to partition B_n into $\left(\begin{bmatrix} n \\ \frac{n}{2} \end{bmatrix} \right)$ chains, for each n. We shall give an inductive proof that such partitions exist, based on a construction discovered by deBruijn, Tengbergen, and Kruyswijk [5] and rediscovered by a number of others.

The essential feature of this construction is that the chains turn out to be *symmetric*: that is, they stretch from a set of size k to a set of size $n - k$, for some k, meeting every intermediate rank. Trivially, if B_n is partitioned into symmetric chains, the number of chains is exactly $\left(\begin{bmatrix} n \\ \frac{n}{2} \end{bmatrix} \right)$.

The construction is as follows: suppose that the subsets of $\{1, 2, \ldots, n - 1\}$ have been partitioned into symmetric chains. For each chain C of length $k \geq 1$, construct a second chain C' by adding n to each set in C. Then construct two new chains in B_n by removing the top of C' and adding it to C. The new chains have lengths $k - 1$ and $k + 1$, and trivially both are symmetric. (If C' becomes empty at this stage it is disregarded.) Applying this procedure to every chain gives a partition of B_n into symmetric chains as desired.

DeBruijn, Tengbergen, and Kruyswijk showed that a minor modification of this procedure works for divisors of an integer as well. That is, the divisors of an integer $N = \prod_{i=1}^{m} p_i^{e_i}$ can be partitioned into $N_k(M_{\vec{e}})$ symmetric chains, where $k = [\Sigma e_i/2]$. (We omit the details.)

Greene and Kleitman [24] and independently Leeb [unpublished] found that the partition constructed above could be described explicitly in the following way:

First associate to each subset $S \subseteq \{1, 2, \ldots, n\}$ a sequence of left and right parentheses, replacing each element of S by a right parenthesis, and each element of the complement of S by a left parenthesis. For example, if $n = 9$ and $S = \{1, 3, 4, 7, 8\}$, we obtain the sequence

```
)  (  )  )  (  (  )  )  (
1  2  3  4  5  6  7  8  9
   ‾‾‾     ‾‾‾‾‾‾‾‾
```

Every sequence of left and right parentheses has a unique "parenthesization" obtained as follows: close all pairs of left and right parentheses which are either adjacent or separated by other such pairs, repeating the process until no further pairing is possible. Note that the remaining unpaired parentheses must necessarily consist of "rights" followed by "lefts".

Now define a partition of B_n by saying that two sets are in the same block if they have the same "parenthesization". From the above remark about unpaired elements, it follows that two sets in the same block must be comparable; hence we have partitioned B_n into chains. In general, a chain in this partition is obtained by starting at the bottom with a set of elements which can be completely paired, and adding the unpaired elements from left to right, one at a time.

For example, the chain in B_9 which contains $S = \{1, 3, 4, 7, 8\}$ consists of the sets $\{3, 7, 8\}$, $\{1, 3, 7, 8\}$, $\{1, 3, 4, 7, 8\}$, $\{1, 3, 4, 7, 8, 9\}$, obtained by adding 1, 4, and 9 in order to $\{3, 7, 8\}$.

It is immediate that the chains constructed in this way are all symmetric, and that the procedure coincides exactly with the one defined inductively by deBruijn, Tengbergen, and Kruyswijk.

Using the above construction, it is possible to obtain the conclusion of Sperner's theorem from slightly weaker hypotheses. It is easy to see that the "unpaired" elements alternate odd-even (since blocks of consecutive paired elements are always even in number); hence the difference of any two members of the same chain is a set which alternates "odd-even". Hence the bound of $\binom{n}{\lceil \frac{n}{2} \rceil}$ remains valid if we exclude only comparable pairs of sets whose difference alternates odd-even. A slightly weaker result (arising from a stronger hypothesis) can be expressed in terms of colorings. If a set X has been colored with two colors—say red and blue—we call the coloring *balanced* if the number of reds differs by at most one from the number of blues.

COROLLARY 3.1: *Suppose that $\{1, 2, \ldots, n\}$ has been given a balanced coloring. If F is a family of subsets containing no comparable pairs of sets whose difference is balanced, then $|F| \leq \binom{n}{\left[\frac{n}{2}\right]}$.*

Corollary 3.1 follows from our previous observation if we renumber the elements so that the coloring is represented by odds and evens.

Corollary 3.1 contrasts with another extension of Sperner's theorem obtained independently by Kleitman [35] and Katona [31].

THEOREM 3.2: *Suppose that the elements of $\{1, 2, \ldots, n\}$ have been given an arbitrary 2-coloring. Let F be a family of sets containing no comparable pair whose difference is monocolored. Then $|F| \leq \binom{n}{\left[\frac{n}{2}\right]}$.*

The proof of Theorem 3.2 is as follows:

Suppose that the coloring partitions $\{1, 2, \ldots, n\}$ into two parts X and Y. We apply our partitioning procedure separately to the lattices B_X and B_Y, producing a collection of symmetric chains for each. Since B_n is isomorphic to the cartesian product $B_X \times B_Y$, we can partition B_n into "symmetric rectangles" by taking all possible products of pairs of chains, one from B_X and the other from B_Y. If F is a family which satisfies the conditions of Theorem 3.2, it is clear that no rectangle contains two members of F in the same row or column. It follows immediately that *in each rectangle* the number of elements of F is bounded by the number of elements of size $\left[\frac{n}{2}\right]$. Since the rectangles partition B_n, F can have at most $\binom{n}{\left[\frac{n}{2}\right]}$ members altogether.

We show next how the methods of this section can be used to derive an upper bound on the total number of antichains in B_n. This is a problem originally posed by Dedekind, and can be rephrased in several ways: for example, it is equivalent to finding a bound for the size of the free distributive lattice on n generators. The argument presented here is due to Hansel [61].

THEOREM 3.3: *The number of antichains in B_n is at most*

$$3^{\binom{n}{\lfloor \frac{n}{2} \rfloor}}.$$

Proof: Antichains in B_n are in one-to-one correspondence with monotone Boolean functions on B_n, that is, order preserving maps from B_n to the set $\{0, 1\}$. We can construct such functions in the following way: take the chains in the deBruijn-Tengbergen-Kruyswijk partition, and beginning with the *smallest chains* first, assign values 0 and 1 to the members of each chain, consistent with the requirements of monotonicity. It can be shown easily that a chain always has at most two unassigned elements (which in fact must be adjacent), once values have been given to the members of all smaller chains. These two elements can be assigned values in at most three ways, and the bound follows immediately.

By a much more difficult argument (based on the same ideas) Kleitman [62] has improved this bound to

$$2^{\binom{n}{\lfloor \frac{n}{2} \rfloor}(1 + O((\log n)/n))}.$$

This result can be interpreted as saying that "most" antichains in B_n are obtained by taking collections of sets of size $\left[\dfrac{n}{2}\right]$.

The method of parenthesizations can be extended readily to lattices of multisets. To each element $(\sigma_1, \sigma_2, \ldots, \sigma_n) \in M_{\vec{e}}$ (where $\vec{e} = (e_1, e_2, \ldots, e_n)$) associate a sequence of Σe_i left and right parentheses, as follows: first σ_1 right parentheses, then $e_1 - \sigma_1$ left, then σ_2 right, $e_2 - \sigma_2$ left, and so forth. Now define two elements σ and τ to be equivalent if the corresponding sequences have the same basic parenthesization. Then all of the previous arguments carry over, and we obtain a partition of $M_{\vec{e}}$ into symmetric chains. This is equivalent to the partition obtained (inductively) by deBruijn, Tengbergen, and Kruyswijk [5].

It would be extremely interesting to obtain a similar explicit construction for lattices of subspaces, but none is known.

Our proof that $M_{\vec{e}}$ can be partitioned into symmetric chains actually proves that the analog of Erdös' theorem for k-families holds as well:

THEOREM 3.4: *If F is a family of multisets $\vec{\sigma} \in M_{\vec{e}}$ which contains no chains of length $k + 1$, then $|F|$ is bounded by the largest sum of k Whitney numbers $N_i(M_{\vec{e}})$.*

A similar result holds for every partially ordered set which can be partitioned into symmetric chains—that is, the maximum size of a k-family is equal to the sum of the k largest Whitney numbers. In such cases, the "largest k" will always be the "middle k" (which of course may not be unique).

We conclude this section by mentioning a result of Greene and Kleitman [23] which shows that—in principle—the maximum size of a k-family in any partially ordered set can always be computed by looking at the right partitions of P into chains.

THEOREM 3.5: *Let P be an arbitrary partially ordered set. Then the maximum size of a k-family in P is equal to the minimum, over all partitions $\mathcal{C} = C_1, C_2, \ldots, C_q$ of P into chains C_i, of the expression*

$$\sum_{i=1}^{q} \min\{|C_i|, k\}.$$

This result reduces to Dilworth's theorem if $k = 1$.

4. THE LYM PROPERTY

We begin this section with a third proof of Sperner's theorem, due independently to Lubell [48], Yamamoto [57], and Meschalkin [50]. This method is much more powerful than those discussed in earlier sections, and a large number of generalizations and extensions are possible. In contrast to previous methods this approach applies simultaneously to lattices of sets, multisets, and subspaces (but again not to partitions).

A *maximal chain* in B_n is a sequence of sets $\Phi = X_0 \subset X_1 \subset \cdots \subset X_n$, where $|X_i| = i$ for each i. There are exactly $n!$ maximal chains in B_n, and exactly $k!(n - k)!$ pass through a given set S of size k. If F is an antichain in B_n, then each maximal chain contains at most one member of F, and so

$$\sum_{S \in F} |S|!(n - |S|)! \leq n!.$$

If we denote by f_k the number of sets in F of size k, this inequality becomes

(*) $$\sum_{k=0}^{n} \frac{f_k}{\binom{n}{k}} \leq 1.$$

But $\binom{n}{k} \leq \binom{n}{\left[\frac{n}{2}\right]}$ for all k, and it follows immediately that

$$|F| = \sum_{k=0}^{n} f_k \leq \binom{n}{\left[\frac{n}{2}\right]}$$

The inequality (*) is actually a stronger statement than Sperner's theorem. We refer to (*) as the *Lubell-Yamamoto-Meschalkin* (LYM) *inequality*. If a partially ordered set P has a rank function, and if

$$\sum_{x \in F} \frac{1}{N_x} = \sum_{k} \frac{f_k}{N_k} \leq 1$$

holds for every antichain $F \subseteq P$, we say that P has the *LYM Property*. Clearly the LYM property always implies the Sperner property.

The above derivation of the LYM inequality uses only the following property of sets: for each set S of size k, the number of maximal chains which pass through S depends only on k and not on S. Hence by a similar argument, we obtain an important sufficient condition for a partially ordered set to have the LYM property.

THEOREM 4.1: *If P is a partially ordered set with a rank function, and if each element of rank k in P is contained in the same number of maximal chains of P (for all k), then P has the LYM property.*

The next observation is due to Baker [2].

COROLLARY 4.2: *If P is "regular" in the sense that for every k, each element of rank k covers the same number of elements of rank $k - 1$ (and dually), then P has the LYM property.*

To prove Corollary 4.2, observe that the number of maximal chains through a given element is completely determined by the "covering numbers", and hence depends only on rank.

In section 2 we proved that the Sperner property holds if the automorphisms of P are transitive on elements of fixed rank. Corollary 4.2 shows that a stronger conclusion (the LYM property) follows from a much weaker assumption (regularity).

COROLLARY 4.3: *For all n and q, $L_n(q)$ has the LYM property (and hence the Sperner property).*

The conditions of theorem 4.1 do not hold for lattices of multisets. However, it turns out that the LYM property remains true, for much deeper reasons to be discussed later (see Corollary 4.12).

Kleitman observed [37] that the hypotheses of Theorem 4.1 can be significantly weakened:

THEOREM 4.4: *Let P be a partially ordered set with a rank function. Suppose that there exists a nonempty collection $\mathcal{C}$ of maximal chains in P (not necessarily distinct) such that, for each k, every element of rank k occurs in the same number of chains in $\mathcal{C}$. Then P has the LYM property.*

If the conditions of Theorem 4.1 hold, we can take $\mathcal{C}$ to be the collection of all maximal chains in P. More generally the argument works as before: each element of rank k must occur in exactly $|\mathcal{C}|/N_k$ members of $\mathcal{C}$. Hence if F is an antichain, we have

$$\Sigma f_k \frac{|\mathcal{C}|}{N_k} \leq |\mathcal{C}|$$

and the LYM inequality follows.

A collection of chains $\mathcal{C}$ which satisfies the hypotheses of Theorem 4.4 will be called a *regular covering of P by chains*. Surprisingly, the existence of such coverings turns out to be equivalent to the LYM property. In fact, both are equivalent to a third hypothesis introduced independently by Graham and Harper [22], called the *normalized matching property*. This property can be described as follows:

If A is any subset of elements of rank k in P, let A^* denote the set of elements of rank $k + 1$ which are related to some element of A. If

$$\frac{|A|}{N_k} \leq \frac{|A^*|}{N_{k+1}}$$

for every k and every such A, then P is said to have the normalized matching property.

For lattices of sets, the statement of normalized matching property is equivalent to the following elementary but useful lemma (apparently due originally to Sperner [56]):

LEMMA 4.5: *Let A be a collection of k-subsets of $\{1, 2, \ldots, n\}$ and let A^* be the collection of $(k + 1)$-subsets which contain members of A. Then*

$$|A| \le \frac{k + 1}{n - k} |A^*|.$$

Kleitman [37] proved the following:

THEOREM 4.6: *For a partially ordered set P with a rank function the following conditions are equivalent*:

(1) *P has the LYM property.*
(2) *P has the normalized matching property.*
(3) *There exists a regular covering of P by chains.*

The implication (3) → (1) has already been observed (Theorem 4.4). We complete the proof of Theorem 4.6 in two steps:

(1) → (2). Suppose that P has the LYM property, and A is a set of elements of rank k. Let P_{k+1} denote the elements in P of rank $k + 1$. Then $A \cup (P_{k+1} - A^*)$ is an antichain of P, so that by the LYM inequality we have

$$\frac{|A|}{N_k} + \frac{|P_{k+1} - A^*|}{N_{k+1}} \le 1.$$

But this inequality trivially implies

$$\frac{|A|}{N_k} \le \frac{|A^*|}{N_{k+1}}.$$

(2) → (3) Assume that the normalized matching property holds, and define $M = \Pi_i N_i$. Define a new partially ordered set P' as follows: for each i, take M/N_i copies of the elements of rank i in P, with each copy of x less than each copy of y if $x < y$ in P. It is trivial to show that the normalized matching property for P implies

the existence of *ordinary* matchings between successive ranks of P', using P. Hall's matching theorem [26]. Putting these matchings together, we obtain a collection of M maximal chains in P which cover each element of rank k exactly M/N_k times. This completes the proof.

If P has the LYM property, there is an important analog of the LYM inequality which holds for *all* subfamilies of P, and permits arbitrary weighting of the elements of P.

THEOREM 4.7: *Let P be a partially ordered set with the LYM property, and let λ be a real-valued function defined on P. For any subset $G \subseteq P$*

$$\sum_{x \in G} \frac{\lambda_x}{N_x} \le \max_{C \in \mathcal{C}} \sum_{y \in C \cap G} \lambda_y.$$

Here $\mathcal{C}$ denotes a regular covering of P by chains. The proof can be expressed easily in probabilistic language, as follows: choose a chain C at random from $\mathcal{C}$ and record the sum $\sum_{y \in C \cap G} \lambda_y$. This defines a random variable on $\mathcal{C}$ whose expected value and maximum value are given, respectively, by the left and right hand sides of the inequality in Theorem 4.7. This proves the theorem.

By taking $\lambda_x = N_x$ in Theorem 4.7, we obtain the following corollary:

COROLLARY 4.8: *If P has the LYM property, and G is any subset of P, then*

$$|G| \le \max_{C \in \mathcal{C}} \left\{ \sum_{x \in C \cap G} N_x \right\}.$$

We mention three examples of applications of Corollary 4.8, to give the reader some idea of its usefulness. The results are stated for lattices of sets, but the obvious analogs hold for any partially ordered set with the LYM property.

THEOREM 4.9: *Let G be a family of subsets of $\{1, 2, \ldots, n\}$.*

(i) (Erdös [15].) *If G contains no chains of length $k + 1$, then $|G|$ is at most the sum of the k largest binomial coefficients $\binom{n}{i}$.*

(ii) (Erdös [15].) *If G contains no two members $A \supseteq B$ with $|A - B| \geq k$, then $|G|$ is at most the sum of the k largest binomial coefficients $\binom{n}{i}$.*

(iii) (Katona [32].) *If G contains no two members $A \supseteq B$ with $|A - B| < k$, then $|G|$ is bounded by the largest sum of the form*

$$\sum_i \binom{n}{a + ki}.$$

Many similar results can be obtained by considering other restrictions on subfamilies. A bound of the form

$$\sum_{k \in S} N_k(P)$$

for some set S of indices can always be obtained in this way, although it will not always be the best possible bound. A best bound is obtained only when the corresponding union of "levels" obeys the restriction in question.

When P does not possess symmetry (or regularity), it is usually difficult to tell whether or not the LYM property holds. However, under certain conditions it can be shown that the LYM property is preserved under direct products. The following result was obtained first by Harper [27] and independently by Hsieh and Kleitman [29]:

THEOREM 4.10: *Let P_1 and P_2 be partially ordered sets, such that*
(i) *both P_1 and P_2 have the LYM property, and*
(ii) *the Whitney numbers of P_1 and P_2 are logarithmically concave (i.e., $N_k^2 \geq N_{k-1} N_{k+1}$ for all k).*
Then both (i) and (ii) hold in the cartesian product $P_1 \times P_2$.

PROOF TECHNIQUES IN THE THEORY OF FINITE SETS 41

In fact, condition (ii) is preserved under products, independently of (i). Note that the Whitney numbers of $P_1 \times P_2$ are obtained from those of P_1 and P_2 by convolution. That is,

$$N_k(P_1 \times P_2) = \sum_{i+j=k} N_i(P_1)N_j(P_2).$$

LEMMA 4.11: *The convolution of two logarithmically concave sequences is logarithmically concave.*

(The proof is not difficult and is omitted; see [27] or [29].)

We proceed now to the proof of Theorem 4.10. The first step is based on an idea which can be regarded as an "LYM-analog" of Theorem 3.2. Let $\mathcal{C}_1$ and $\mathcal{C}_2$ be regular coverings of P_1 and P_2 by chains. By taking all possible products $C_1 \times C_2$ of chains $C_1 \in \mathcal{C}_1$ and $C_2 \in \mathcal{C}_2$, we can cover $P_1 \times P_2$ by "rectangles", and this covering is "regular" in the sense that the number of rectangles containing an element $(x, y) \in P_1 \times P_2$ depends only on $r(x)$ and $r(y)$. By the same argument as that used to prove Theorem 4.7, we obtain the following inequality, for any subset $G \subseteq P_1 \times P_2$ and any weight function λ:

$$\sum_{\substack{x \in G \\ x=(x_1, x_2)}} \frac{\lambda_x}{N_{x_1} N_{x_2}} \leq \max_{\substack{C_1 \in \mathcal{C}_1 \\ C_2 \in \mathcal{C}_2}} \{ \sum_{\substack{y \in G \\ y \in C_1 \times C_2}} \lambda_y \}.$$

Now suppose that G is an antichain of $P_1 \times P_2$. We wish to prove that the LYM inequality holds for G. For each $x = (x_1, x_2) \in P_1 \times P_2$, take

$$\lambda_x = \frac{N_{x_1} N_{x_2}}{N_x}$$

in the previous inequality. Then

$$\sum_{x \in G} \frac{1}{N_x} \leq 1$$

follows if we can show that

$$\sum_{y \in G \cap (C_1 \times C_2)} \frac{N_{y_1} N_{y_2}}{N_y} \leq 1$$

for every pair of chains $C_1 \in \mathcal{C}_1$, $C_2 \in \mathcal{C}_2$. This inequality can be proved by showing that the elements of smallest rank in $G \cap (C_1 \times C_2)$ can be shifted up one level, without decreasing the sum on the left. Hence repeating this operation allows us to concentrate the elements at a single level, where the inequality is trivial. The contribution of the elements of smallest rank in $G \cap (C_1 \times C_2)$ will consist of "connected" blocks of the form

$$\frac{1}{N_k(P_1 \times P_2)} \sum_{i=a}^{b} N_i(P_1) N_{k-i}(P_2).$$

It suffices to show that raising each of these blocks separately does not decrease the sum, since the blocks do not interact with each other. Writing

$$S_k[a, b] = \sum_{i=a}^{b} N_i(P_1) N_{k-i}(P_2),$$

we must prove that

$$\frac{S_k[a, b]}{S_k[0, \infty]} \leq \frac{S_{k+1}[a, b+1]}{S_{k+1}[0, \infty]}.$$

But this can be proved by writing each side as a telescoping product and verifying the inequality

$$\frac{S_k[u, v]}{S_k[u, v+1]} \leq \frac{S_{k+1}[u, v+1]}{S_{k+1}[u, v+2]},$$

which is a straightforward consequence of the logarithmic concavity of the N_i's. (We omit the details.) This completes the proof.

The final steps in the above argument amount to proving a form of the LYM inequality for certain "weighted" sums over rectangles. More general weighted inequalities of this type have been obtained in [27], [37], and [46].

COROLLARY 4.12: *Every finite lattice of multisets (or lattice of divisors of an integer) has the LYM property.*

Thus, of the four classes of lattices mentioned in section 1, the first three (sets, multisets, and subspaces) have the LYM property. On the other hand, Spencer [54] has shown that for sufficiently large n, the lattice of partitions of $\{1, 2, \ldots, n\}$ does *not* have the LYM property. However, many consequences of the LYM property remain open (and important) questions.

5. GENERALIZED LYM INEQUALITIES

In this section, we will consider several other problems which can be solved by variations of the LYM approach. We will use more general systems of sets instead of maximal chains to derive upper bounds on the size of "antisystems", using arguments similar to those described in the last section. If there is suitable symmetry, one can always obtain inequalities of the LYM type.

An *ordered set-system* on $\{1, 2, \ldots, n\}$ is a sequence of sets $\vec{\alpha} = (A_1, A_2, \ldots, A_p)$, where each A_i is a subset of $\{1, 2, \ldots, n\}$. Let $\mathcal{S}(\vec{\alpha})$ denote the collection of all ordered set systems $\vec{\beta}$ which can be obtained from $\vec{\alpha}$ by permuting the elements $1, 2, \ldots, n$. The arguments in this section are all based on the following general result:

THEOREM 5.1: *Let $\vec{\alpha} = (A_1, A_2, \ldots, A_p)$ be an ordered set system on $\{1, 2, \ldots, n\}$, with $|A_i| = \alpha_i$, $1 \le i \le p$. Let F be a family of subsets of $\{1, 2, \ldots, n\}$ such that F has at most k members in common with each $\vec{\beta} \in \mathcal{S}(\vec{\alpha})$. Let f_j denote the number of sets in F of size j. Then*

$$\sum_{i=1}^{p} \frac{f_{\alpha_i}}{\binom{n}{\alpha_i}} \le k.$$

The proof is by exactly the same argument as that used to prove Theorem 4.7.

If $\vec{\alpha}$ represents a maximal chain of sets in B_n (that is, if $\Phi = A_1 \subset A_2 \subset \cdots \subset A_{n+1}$) and F is an antichain, then Theorem 5.1 yields the ordinary LYM inequality for families of sets.

By similar reasoning, it is also possible to obtain an analog of Theorem 4.7, which permits weighting of sets and applies to arbitrary subfamilies $G \subseteq B_n$.

COROLLARY 5.2: *Let G be any family of subsets of $\{1, 2, \ldots, n\}$, and let λ_i be an arbitrary weight assigned to sets of size i ($1 \leq i \leq n$). If $\vec{\alpha}$ is any ordered set system then (with the notation defined above)*

$$\sum_{i=1}^{p} \frac{\lambda_{\alpha_i} g_{\alpha_i}}{\binom{n}{\alpha_i}} \leq \max_{\beta \in \mathcal{S}(\vec{\alpha})} \left(\sum_{A \in G \cap \vec{\beta}} \lambda_{|A|} \right).$$

As a first application of Theorem 5.1, we will give a short proof of the following theorem due to Erdös, Ko, and Rado [17]. The proof is a slight modification of one originally given by Katona [33].

THEOREM 5.3: *Let $k \leq n/2$ and let F be a family of k-subsets of $\{1, 2, \ldots, n\}$, no two members of which are disjoint. Then $|F| \leq \binom{n-1}{k-1}$.*

Proof: Let $\vec{\alpha} = (A_1, A_2, \ldots, A_n)$ be the set system obtained by arranging the numbers $1, 2, \ldots, n$ in a circle and taking the A_i's to be all consecutive segments of length k. We claim that no sequence $\vec{\beta} \in \mathcal{S}(\vec{\alpha})$ contains more than k members of F. By symmetry, this follows if we can prove that $\vec{\alpha}$ contains at most k members of F. But if A_i denotes the segment beginning with i (mod n), and F contains A_1, then F contains at most one set from each of the pairs $\{A_{i-k}, A_i\}$ for $i = 2, 3, \ldots, k$, and no others. Hence F has at most k sets in common with α. By Theorem 5.1,

$$\frac{nf_k}{\binom{n}{k}} = \frac{n|F|}{\binom{n}{k}} \le k$$

and the result follows.

An even easier argument, similar to the above, can be obtained in the special case when k divides n. Take $\vec{\alpha}$ to be any partition of $\{1, 2, \ldots, n\}$ into n/k blocks of size k so that $S(\vec{\alpha})$ is the set of all partitions of this type. Trivially, each partition contains at most one member of F, and Theorem 5.1 yields the inequality

$$\frac{n}{k} \cdot \frac{|F|}{\binom{n}{k}} \le 1$$

from which the result follows as before.

In their original paper, Erdös, Ko, and Rado derived Theorem 5.3 from a more general hypothesis: the sets in F are assumed to have size *at most* k, and the same bound $\binom{n-1}{k-1}$ follows. By a slight change in the above proof, we can obtain this result as a consequence of an even stronger statement, which is an LYM inequality for "Erdös-Ko-Rado families".

THEOREM 5.4: *Let F be an antichain of subsets of $\{1, 2, \ldots, n\}$, each of size at most $n/2$, such that no two members of F are disjoint. If f_i denotes the number of sets in F of size i, then*

$$\sum_{i=1}^{n/2} \frac{f_i}{\binom{n-1}{i-1}} \le 1.$$

Proof: Instead of taking $\vec{\alpha}$ to be the sequence of consecutive k-segments on a circle, let $\vec{\alpha}$ be the collection of consecutive segments of *all* lengths, in some arbitrary order. In addition, assign a weight $\lambda_j = 1/j$ to sets of size j, and apply Corollary 5.2. Since there are n segments of length α_j for all $\alpha_j > 0$, the left hand side of the inequality in Corollary 5.2 becomes

$$\sum_{j=1}^{n/2} \frac{n}{j} \frac{f_j}{\binom{n}{j}} = \sum_{j=1}^{n/2} \frac{f_j}{\binom{n-1}{j-1}}.$$

Thus to prove the result, we must prove that the right hand side of the inequality in Corollary 5.2 is at most 1. Equivalently, we must show that

$$\sum_{A \in F \cap \vec{\alpha}} \frac{1}{|A|} \le 1.$$

But this can be proved as follows: suppose that the smallest consecutive segment occurring in $F \cap \vec{\alpha}$ has length k. Then, as in the proof of Theorem 5.3, $F \cap \vec{\alpha}$ has at most k members. Since $1/|A| \le 1/k$ for all $A \in F \cap \vec{\alpha}$, the inequality follows.

The above argument is essentially due to Bollobas [3], but was discovered independently by Greene, Kleitman, and Katona [25]. Theorem 5.4 suggests that the parameters f_j corresponding to an Erdös-Ko-Rado family behave like those of a Sperner family on a set of size $n - 1$. In [25] the latter authors strengthened Theorem 5.4 by proving that the f_j's are completely characterized by this property. (See Theorem 8.12.)

Bollobas [3] used Theorem 5.4, to derive a result about a special class of antichains:

COROLLARY 5.5: *Let F be an antichain of subsets of* $\{1, 2, \ldots, n\}$ *such that for every* $A \in F$, *the complement of A is also in F. Then*

$$\sum_{A \in F} \frac{1}{\mu(A)} \le 2,$$

where $\mu(A) = \min \left\{ \binom{n-1}{|A|-1}, \binom{n-1}{n-|A|-1} \right\}.$

To prove Corollary 5.5, observe that an antichain F satisfies the given hypothesis if and only if it is obtained by taking the union of an Erdös-Ko-Rado family (whose members all have size $\leq n/2$) and the family of its complements. The inequality now follows from two applications of Theorem 5.4.

Bollobas [3] also observed that Corollary 5.5 can be used to prove the following result due independently to Kleitman and Spencer [44] and Schönheim [53]:

COROLLARY 5.6: *Let F be an antichain of subsets of* $\{1, 2, \ldots, n\}$ *such that for all* $A, B \in F$, $A \cap B \neq \emptyset$ *and* $A \cup B \neq \{1, 2, \ldots, n\}$. *Then*

$$|F| \leq \binom{n-1}{\lceil \frac{n}{2} \rceil - 1}.$$

To prove Corollary 5.6, let $\bar{F}$ denote the family of all sets which are complements of sets in F. The given conditions imply that $F \cup \bar{F}$ is an antichain, so we may apply Corollary 5.5. Since $\mu(A) \leq \binom{n-1}{\lceil \frac{n}{2} \rceil - 1}$ for every set A, it follows that

$$\frac{2|F|}{\binom{n-1}{\lceil \frac{n}{2} \rceil - 1}} \leq \sum_{A \in F \cup \bar{F}} \frac{1}{\mu(A)} \leq 2.$$

The inequality in Corollary 5.6 is best possible, since we can construct an extremal family satisfying the desired conditions by taking F to be all $[n/2]$-sets containing a fixed element.

Hsieh [28] proved that an analog of the Erdös-Ko-Rado theorem holds for subspaces of a finite vector space, except that his proof works only when $k < n/2$, leaving the case $k = n/2$ unsettled. We shall show below that Katona's proof can be modified to cover this case. (In fact, it works whenever k divides n.) Combining these results, we can state

THEOREM 5.7: *Let F be a family of k-dimensional subspaces of* a vector space of dimension n over $\mathrm{GF}(q)$, with $k \leq n/2$. If no two members of F have intersection $\{0\}$, then $|F| \leq \begin{bmatrix} n-1 \\ k-1 \end{bmatrix}_q$.

We will give the proof only for the case $k|n$.

LEMMA 5.8: *Let V be an n-dimensional vector space over* $\mathrm{GF}(q)$, *and let k be an integer which divides n. Then* $V-\{0\}$ *can be partitioned into sets of the form* $K-\{0\}$, *where K denotes a k-dimensional subspace of V.*

To prove Lemma 5.8 take V to be the finite field of order q^n. If K_0 is the subfield of order q^k, then the cosets of $K_0-\{0\}$ in the multiplicative group of V have the desired property.

To prove Theorem 5.7 when k divides n, we take a fixed partition of V as guaranteed by Lemma 5.8, and "symmetrize" by taking all partitions obtainable from the first by linear transformations. Then the proof of Theorem 5.3 (or more precisely, the remark following it) remains valid and Theorem 5.7 follows immediately.

We have not been able to show that the analog of a "k-fold covering" exists when $k \nmid n$, which seems to be more difficult for vector spaces than it is for sets. (In this case what is needed is a list of $(q^n - 1)$ k-subspaces which meets an intersecting family at most $q^k - 1$ times.) If such coverings could be constructed another proof of Hsieh's theorem would follow.

As our second example, we will prove the following theorem due to Kleitman [36]:

THEOREM 5.9: *Let* $n = 3k + 1$, *and let G be a family of subsets of* $\{1, 2, \ldots, n\}$ *which contains no two disjoint sets and their union. Then*

$$|G| \leq \sum_{i=k+1}^{2k+1} \binom{n}{i}.$$

Thus a family of maximum size is obtained by taking G to be the collection of all subsets A with $k + 1 \leq |A| \leq 2k + 1$. The proof rests on the following lemma:

LEMMA 5.10: *Let $\bar{g}_i$ denote the number of i-sets which are not in G, and suppose $a + b + c = n$. Then*

$$\frac{\bar{g}_a}{\binom{n}{a}} + \frac{\bar{g}_b}{\binom{n}{b}} + \frac{\bar{g}_c}{\binom{n}{c}} + \frac{\bar{g}_{a+b}}{\binom{n}{a+b}} + \frac{\bar{g}_{a+c}}{\binom{n}{a+c}} + \frac{\bar{g}_{b+c}}{\binom{n}{b+c}} \geq 2.$$

To prove the Lemma, let $\vec{\alpha}$ be the ordered set system $(A, B, C, A \cup B, A \cup C, B \cup C)$, where $A \cup B \cup C$ is a partition of $\{1, 2, \ldots, n\}$ and $|A| = a$, $|B| = b$, and $|C| = c$. It follows from the conditions on G that each sequence $\vec{\beta} \in \mathcal{S}(\vec{\alpha})$ contains at least two sets which are *not* in G. The lemma follows immediately from Theorem 5.1.

To prove Theorem 5.9, suppose that $a \leq b \leq c$ and $a + b + c = n$. By Lemma 5.10,

$$\bar{g}_a + \bar{g}_{n-a} + \frac{\binom{n}{a}}{\binom{n}{b}}(\bar{g}_b + \bar{g}_{n-b}) + \frac{\binom{n}{a}}{\binom{n}{c}}(\bar{g}_c + \bar{g}_{n-c}) \geq 2\binom{n}{a}.$$

For each $a < k$, choose $b = \left\lceil \dfrac{n-a}{2} \right\rceil - 1$ and $c = n - a - b$. Then if we sum these inequalities for all $a < k$, and add the single inequality

$$\bar{g}_k + \bar{g}_{n-k} + \frac{1}{2}\frac{\binom{n}{k}}{\binom{n}{k+1}}(\bar{g}_{k+1} + \bar{g}_{n-k-1}) \geq \binom{n}{k},$$

it can be shown by a simple calculation that the left hand side is at most

$$\sum_{i=0}^{n} \bar{g}_i = 2^n - |G|$$

while the right hand side is exactly

$$\sum_{i \notin [k+1, 2k+1]} \binom{n}{i}.$$

Hence $|G| \leq \sum_{i=k+1}^{2k+1} \binom{n}{i}$ as desired.

Modifications in the above argument can be made to obtain analogous results for collections of subspaces. We omit the details.

6. LINEAR PROGRAMMING TECHNIQUES

In the last two sections, we have described techniques for obtaining certain kinds of linear inequalities which antichains and other families must obey. While these inequalities often have immediate consequences (e.g., Sperner's Theorem, the Erdös-Ko-Rado Theorem), it is sometimes possible to obtain deeper results by using the techniques of linear programming. In this section, we will explore several examples which illustrate this approach.

We begin by proving two closely related theorems, the first due to Kleitman [60], and the second due to Kleitman and Milner [59].

THEOREM 6.1: *Let F be an antichain in B_n, with $|F| \geq \binom{n}{k}$, $k \leq n/2$. If F^* denotes the order ideal generated by F (i.e., the family of sets contained in some member of F), then*

(i) $\qquad |F^*| \geq \sum_{i \leq k} \binom{n}{i},\qquad$ (Kleitman [60])

(ii) $$\frac{1}{|F|} \sum_{A \in F} |A| \geq k. \text{ (Kleitman, Milner [59])}$$

In other words, if $|F| \geq \binom{n}{k}$, then both the size of F^* and the average size of the members of F are at least as big as they would be if F were concentrated at level k.

The method of proof actually applies to any partially ordered set P with the LYM property whose Whitney numbers are logarithmically concave. We shall state and prove the result in this form:

THEOREM 6.2: *Let P be a partially ordered set with the LYM property, such that $(N_i(P))^2 \geq N_{i-1}(P)N_{i+1}(P)$ for each i. Let k be any integer such that $k \leq m$, where m denotes the index i for which $N_i(P)$ is maximum. Let $F \subseteq P$ be an antichain satisfying $|F| \geq N_k(P)$, and let F^* denote the order ideal generated by F. Then*

(i) $$|F^*| \geq \sum_{i \leq k} N_i(P),$$

(ii) $$\frac{1}{|F|} \sum_{x \in F} r(x) \geq k.$$

Proof: For each i, let $x_i = f_i/N_i$, where f_i denotes the number of elements in F having rank i. Since P has the LYM property, we have

$$\Sigma x_i \leq 1.$$

The assumption $|F| \geq N_k$ gives a second linear constraint:

$$\Sigma x_i N_i \geq N_k.$$

Since P has the LYM property, we can assume that $x_i = 0$ for $i > m$, since otherwise the top levels of F can be lowered without reducing $|F|$ (using normalized matching).

To prove (i), we will relate the problem of minimizing $|F^*|$ to the problem (not equivalent, but sufficient for present purposes) of minimizing a certain linear function. To this end, let f_i^* denote the number of elements in F^* of rank i, for each j, and let $x_i^* = f_i^*/N_i$. Observe that for each i we can construct an antichain in P by taking all elements in F of rank $i \geq j$, together with all elements of rank j which are *not* in F^*. Hence, by the LYM inequality,

$$(1 - x_j^*) + \sum_{i \geq j} x_i \leq 1$$

which implies

$$\sum_{i \geq j} x_i \leq x_j^*$$

for each j. Since $|F^*| = \sum x_j^* N_j$, it follows that

$$|F^*| \geq \sum_j N_j \sum_{i \geq j} x_i = \sum_i x_i \{\sum_{j \leq i} N_j\}.$$

Now consider the linear program

$$\sum_{i=0}^{m} - x_i \geq -1$$

$$\sum_{i=0}^{m} x_i N_i \geq N_k, \quad x_i \geq 0;$$

$$\textit{minimize}: \sum_{i=0}^{m} x_i \overline{N_i},$$

where $\overline{N_i} = \sum_{j \leq i} N_j$. The dual program is:

$$-\alpha + \beta N_i \leq \overline{N_i}, \quad i = 0, 1, 2, \ldots, n,$$

$$\alpha, \beta \geq 0;$$

$$\textit{maximize}: -\alpha + \beta N_k.$$

In order to prove (i), it is sufficient to find values of α and β, satisfying the conditions of the dual program, such that $-\alpha + \beta N_k = \overline{N_k}$. By duality, $\overline{N_k}$ must then be a lower bound for the original problem. To find such values, we set $-\alpha + \beta N_k = \overline{N_k}$ and eliminate α, which leads to the conditions

$$\beta N_k \geq \overline{N_k}, \quad \beta(N_i - N_k) \leq \overline{N_i} - \overline{N_k},$$

for $i = 0, 1, \ldots, n$. This is equivalent to the conditions

$$\frac{\overline{N_k} - \overline{N_j}}{N_k - N_j} \leq \beta \leq \frac{\overline{N_i} - \overline{N_k}}{N_i - N_k}$$

for all i and j such that $-1 \leq j < k < i \leq m$. Using the logarithmic concavity of the N_i's, one can easily show that

$$\frac{\overline{N_k} - \overline{N_j}}{N_k - N_j} \leq \frac{N_k}{N_k - N_{k-1}} \leq \frac{N_{k+1}}{N_{k+1} - N_k} \leq \frac{\overline{N_i} - \overline{N_k}}{N_i - N_k},$$

which proves that suitable values for α and β exist. This completes the proof of part (i).*

To prove part (ii), we can assume that $|F| = N_k$ (throwing away the highest elements if necessary). Then the average rank of members of F is bounded from below by the solution to the linear program

$$\sum_{i=0}^{m} -x_i \geq -1,$$

$$\sum_{i=0}^{m} N_i x_i \geq N_k, \quad x_i \geq 0;$$

$$\text{minimize: } \sum_{i=0}^{m} \frac{iN_i}{N_k} x_i.$$

---

*This argument was communicated to the authors by A. Odlyzko.

The dual program is

$$-\alpha + \beta N_i \leq \frac{iN_i}{N_k}, i = 0, 1, \ldots, m,$$

$$\alpha, \beta, \geq 0;$$

maximize: $-\alpha + \beta N_k$.

As before, we seek to find values of α and β satisfying the dual constraints, such that $-\alpha + \beta N_k = k$. Solving and eliminating α leads to the conditions

$$\beta N_k \geq k, \quad \beta(N_i - N_k) \leq \frac{iN_i}{N_k} - k,$$

for $i = 0, 1, \ldots, m$. These conditions can be satisfied if and only if

$$\frac{kN_k - jN_j}{N_k - N_j} \leq \frac{iN_i - kN_k}{N_i - N_k}$$

for all j, k such that $j < k < i \leq m$, which in turn can be written as

$$\frac{1}{N_k} \leq \frac{(i-k)}{(i-j)} \frac{1}{N_j} + \frac{(k-j)}{(i-j)} \frac{1}{N_i}$$

for $j < k < i \leq m$. The proof of part (ii) can now be completed by the application of the following easy lemma, whose proof we omit:

LEMMA 6.3: *If* $\alpha_0, \alpha_1, \alpha_2, \ldots,$ *is any sequence which is logarithmically concave, then the sequence* $\alpha_0^{-1}, \alpha_1^{-1}, \alpha_2^{-1}, \ldots,$ *is convex (in the usual sense).*

As our last example, we will prove the following theorem due to Kleitman [41]:

THEOREM 6.4: *Let F be a family of sets in B_n which contains no three distinct members A, B, C, with $A \cup B = C$. Then*

$$|F| \leq \left(\begin{bmatrix} n \\ \frac{n}{2} \end{bmatrix} \right) (1 + \frac{1}{n^{1/2}}).$$

Proof: Let f_k denote the number of elements in F of rank k. Fix an integer $j \leq n/2$ and consider the collection $\mathcal{C}_{(j)}$ of maximal chains between rank j and rank $2j$. Let S be a member of F of rank $k \geq j$. We claim the following: among all of the chains in $\mathcal{C}_{(j)}$ passing through S, at least a proportion j/k contain no smaller member of F. This follows immediately from the Erdös-Ko-Rado Theorem (viewed upside down): if we add all of the rank j elements below S which are dominated by members of F, the proportion of chains which meet smaller members of F can only increase. But these sets form a (dual) Erdös-Ko-Rado family of j-subsets of S, whose size must be bounded by $\binom{k-1}{k-j-1}$. Hence the proportion of chains through S which meet smaller members of F must be at least

$$\binom{k-1}{k-j-1} \bigg/ \binom{k}{j} = \frac{k-j}{k}$$

as desired. If we count, for each $S \in F$, the number of chains through S which contain no smaller elements of F, the result is clearly bounded by the total number of chains in $\mathcal{C}_{(j)}$. This implies

$$\sum_{k=j}^{2j} \frac{f_k}{\binom{n}{k}} \frac{j}{k} \leq 1$$

for each $j = 1, 2, \ldots, [n/2]$.

Next consider the linear program whose object is to maximize the linear function

$$\sum_{k=0}^{n} f_k$$

subject to these constraints, together with the conditions $f_k \geq 0$, $k = 0, 1, \ldots, n$. The dual program is to maximize

$$\sum_{k=1}^{[n/2]} \alpha_k$$

subject to the constraints

$$\sum_{k=j/2}^{\min\{j,\, [\frac{n}{2}]\}} \frac{k\, \alpha_k}{j \binom{n}{j}} \geq 1,$$

$j = 1, 2, \ldots, n$, with $\alpha_k \geq 0$, $k = 1, 2, \ldots, \left[\dfrac{n}{2}\right]$.

The values

$$\alpha_{2k+1} = \binom{n}{2k+1} - \binom{n}{2k} + \frac{1}{2k+1}\binom{n}{2k} + \alpha_k,$$

$$\alpha_{2k} = \binom{n}{2k} - \binom{n}{2k-1} + \frac{1}{2k}\binom{n}{2k},$$

satisfy the dual constraints and yield as objective function

$$\binom{n}{[\frac{n}{2}]} + \frac{2^n}{n+1} + O\left(\binom{n}{[\frac{n}{4}]}\right)$$

which implies that the latter is an upper bound for the solution to the original problem. This bound is asymptotic to the one stated in Theorem 6.4.

We note that this bound is only asymptotic and may not be best possible. One can construct examples which satisfy the conditions of the theorem and attain the value $\binom{n}{[\frac{n}{2}]}\left(1 + \dfrac{1}{n}\right)$ in the following way: let $n = 2^\alpha - 1$, and consider the Hamming code H_α of order α. We can think of H_α as a family of n-sets with the property that every pair of sets differ in at least three places. Moreover each of the cosets of H_α can also be thought of as a family with this property. Since $|H_\alpha| = 2^{2^\alpha - \alpha - 1} \sim 2^n/n$, at least one of the cosets must contain a proportion $1/n$ of the $\lceil \frac{n}{2} \rceil$-sets. If we take these $\lceil \frac{n}{2} \rceil$-sets, together with *all* of the $\lceil \frac{n}{2} \rceil + 1$-sets, we obtain a family of size at least $\binom{n}{\lceil \frac{n}{2} \rceil} \left(1 + \frac{1}{n}\right)$ with the desired properties.

7. INTERSECTIONS OF ORDER IDEALS

If P is any partially ordered set, an *order ideal* of P is a subset $K \subseteq P$ such that whenever $x \in K$ and $y \leq x$, then $y \in K$. If $P = B_n$, an order ideal is sometimes called a *simplicial complex*. This section is based on the following elementary but useful fact:

THEOREM 7.1: (Kleitman [38]) *Suppose that F and G are order ideals in B_n. Then*

$$\frac{|F|}{2^n} \cdot \frac{|G|}{2^n} \leq \frac{|F \cap G|}{2^n}.$$

In other words, the proportion of sets in $F \cap G$ is at least as large as the product of the proportions of sets in F and G.[†]

We can turn one of the order ideals upside down by taking complements and derive the following from Theorem 7.1 as an immediate corollary:

[†] Or in probabilistic language, if A is a set chosen at random from B_n, then Prob $\{A \in F\} \leq$ Prob $\{A \in F | A \in G\}$.

COROLLARY 7.2: *Suppose that F is an order ideal in B_n, and G is an order ideal in the dual of B_n. Then*

$$\frac{|F|}{2^n} \cdot \frac{|G|}{2^n} \geq \frac{|F \cap G|}{2^n}.$$

Before proving theorem 7.1, we mention a typical application:

Suppose that $F \subseteq B_n$ is a family of sets with the property that no two members are disjoint (no restriction on rank or comparability is assumed). It is trivial to see that F has at most 2^{n-1} members, since no set can occur together with its complement. The same bound holds for families $G \subseteq B_n$ with the property that no two sets cover all of the points. Daykin and Lovasz [11] proved that a family which satisfies *both* of these conditions can have at most 2^{n-2} members:

THEOREM 7.3: *Let H be a family of subsets of $\{1, 2, \ldots, n\}$ with the property that for all $A, B \in H$, $A \cap B \neq \emptyset$ and $A \cup B \neq \{1, 2, \ldots, n\}$. Then $|H| \leq 2^{n-2}$.*

The upper bound is achieved by taking all subsets which miss one point and contain another.

We can derive Theorem 7.3 from Corollary 7.2 as follows: Let H satisfy the given conditions, and define F to be the order ideal generated by H, and G to be the dual order ideal generated by H. Then no two members of G are disjoint, and no two members of F have union equal to $\{1, 2, \ldots, n\}$. Hence by Corollary 7.2,

$$|H| \leq |F \cap G| \leq \frac{|F| \cdot |G|}{2^n} \leq 2^{n-2}.$$

The statement of Theorem 7.1 is also valid for divisors of any integer, and we shall give a proof of the result in this form:

THEOREM 7.4: *Let $N = \prod_{i=0}^{m} p_i^{e_i}$ be a positive integer, whose prime* *decomposition is as shown. Let* $\delta = \prod_{i=0}^{m} (1 + e_i)$ *denote the total number of divisors of N. If F and G are order ideals of divisors of N, then*

$$\frac{|F|}{\delta} \cdot \frac{|G|}{\delta} \le \frac{|F \cap G|}{\delta}.$$

The proof is by induction on m, the number of distinct prime divisors of N. Fix a prime p_0, and let F_i denote the subset of F consisting of those members in which p_0 occurs to the ith power ($0 \le i \le e_0$). Define G_i and $(F \cap G)_i$ similarly. Let $\delta' = \prod_{i=1}^{m} (1 + e_i)$ denote the total number of divisors of $N/p_0^{e_0}$. If $0 \le i \le j \le e_0$ it is not difficult to show that $|F_i| \, |G_j| \le \delta' \cdot |(F \cap G)_i|$ (by removing all occurrences of p_0 and applying the inductive hypothesis). The result now follows immediately from the following lemma (we omit the details):

LEMMA 7.5: *Let* $(x_0, x_1, \ldots, x_q)$, $(y_0, y_1, \ldots, y_q)$ *and* $(z_0, z_1, \ldots, z_q)$ *be sequences of nonnegative real numbers, such that for every* $i, j \le q$, $x_i y_j \le z_{\min(i,j)}$. Then

$$(\Sigma x_i)(\Sigma y_j) \le (q + 1)(\Sigma z_k).$$

Lemma 7.5 in turn can be derived from the following elementary fact, which is sometimes known as Chebyschev's inequality:

LEMMA 7.6: *Let* $\alpha_0 \le \alpha_1 \le \cdots \le \alpha_q$ *and* $\beta_0 \le \beta_1 \le \cdots \le \beta_q$ *be sequences of nonnegative real numbers. Then*

$$(\Sigma \alpha_i)(\Sigma \beta_i) \le (q + 1)(\Sigma \alpha_i \beta_i).$$

Surprisingly, the analog of Theorem 7.4 fails to hold for subspaces of a finite vector space. Consider a vector space V of dimension 4 over $GF(q)$. Let F consist of all subspaces of dimension 0 and 1, together with half of the subspaces of dimension 2.

Let G consist of all subspaces of dimension 0 and 1, plus the other half of the 2-spaces. If δ denotes the total number of subspaces, then

$$\frac{|F|}{\delta} = \frac{|G|}{\delta} = \frac{1}{2}.$$

On the other hand, $|F \cap G| = 1 + (q^4 - 1)/(q - 1)$, while $\delta = 2(1 + (q^4 - 1)/(q - 1)) + (q^2 + 1)(q^2 + q + 1)$. Hence

$$\frac{|F \cap G|}{\delta} \to 0$$

as $q \to \infty$, which shows that the inequality in Theorem 7.4 cannot be valid.

We conclude this section by giving another application of Corollary 7.2.

THEOREM 7.7: (Kleitman [38]). *If $F_1, F_2, \ldots, F_k$ are disjoint collections of subsets of $\{1, 2, \ldots, n\}$ such that $A \cap B \neq \emptyset$ for all $A, B \in F_i$, $i = 1, 2, \ldots, k$, then*

$$\left| \bigcup_{i=1}^{k} F_i \right| \leq 2^n - 2^{n-k}.$$

The bound is achieved by taking F_1 to be the collection of all sets which contain 1, F_2 to be sets which contain 2 but not 1, F_3 to be the sets which contain 3 but not 2 or 1, and so forth.

The proof of Theorem 7.7 is by induction on k. For $k = 1$, the result is trivial. Moreover, it is not hard to see that if F_i is a *maximal* family satisfying $A \cap B \neq \emptyset$ for all $A, B \in F_i$, then $|F_i| = 2^{n-1}$. If $k > 1$, let U denote the union of the families $F_1, F_2, \ldots, F_{k-1}$. Assuming that the union of all F_i's is as large as possible it follows that U must be a dual order ideal. By the inductive hypothesis, $|U| = 2^n - 2^{n-k+1}$. Extend F_k to a maximal family F_k' (satisfying the above condition with $k = 1$), and let L denote the collection of sets which are not in F_k'. Then L is an order ideal, and $|L| = 2^{n-1}$. We have

$$\left|\bigcup_{i=1}^{k} F_i\right| = |U \cup F_k| \leq |U \cap L| + |F_k'|$$

$$\leq \frac{|U||L|}{2^n} + |F_k'| \quad \text{(by Corollary 7.2)}$$

$$\leq 2^n - 2^{n-k}.$$

8. CANONICAL FORMS

One of the most versatile theorems of extremal set theory was proved by Kruskal [45] and later independently by Katona [34] (although related results were obtained much earlier by Macaulay [49]). The Kruskal-Katona theorem answers the following question:

If F is a family of k-sets, let ∂F denote the family of $(k-1)$-sets which are subsets of members of F. How small can $|\partial F|$ be, given $|F|$?

A crude lower bound on $|\partial F|$ can be obtained from the trivial Lemma 4.5 proved earlier (which is essentially the normalized matching property for sets):

$$|\partial F| \geq \frac{k}{n-k+1} |F|.$$

Kruskal and Katona obtained a much more precise statement, which gives a best possible lower bound on $|\partial F|$:

THEOREM 8.1: *Let F be a family of k-sets. If*

(**) $$|F| = \binom{a_k}{k} + \binom{a_{k-1}}{k-1} + \cdots + \binom{a_i}{i},$$

$$a_k > a_{k-1} > \cdots > a_i \geq i > 0,$$

then

$$|\partial F| \geq \binom{a_k}{k-1} + \binom{a_{k-1}}{k-2} + \cdots + \binom{a_i}{i-1}.$$

It is well known that every positive integer can be expressed uniquely in the form (**). This representation is known as the *k-binomial expansion* of an integer, and its existence can be proved easily from elementary properties of binomial coefficients (see [45]). To find such a representation of $|F|$, choose a_k to be as large as possible with $\binom{a_k}{k} \leq |F|$. Then subtract and repeat with $k-1$, $k-2$, and so forth.

We will defer a proof of Theorem 8.1 until the end of this section.

If the numbers associated with the Kruskal-Katona Theorem seem mysterious at first glance the following observation should help to explain their significance. Consider the set S_k of all infinite sequences $(x_0, x_1, \ldots,)$ of zeros and ones, with exactly k ones in each sequence. We introduce a linear ordering on S_k by ordering sequences with respect to the *last* position in which they differ (reverse lexicographic ordering). Denote the elements of S_k by σ_0, $\sigma_1, \sigma_2, \ldots$.

LEMMA 8.2: *Let $\sigma_m = (x_0, x_1, \ldots,)$ denote the m-th member of S_k, and suppose that σ_m has ones in positions with indices $a_1 < a_2 < \cdots < a_k$. Then*

$$m = \binom{a_k}{k} + \binom{a_{k-1}}{k-1} + \cdots + \binom{a_1}{1}$$

(where by convention a term is zero if its numerator is less than its denominator).

For example, $(0, 0, 1, 0, 1, 1, 0, \ldots,)$ is the 18th member of S_3, since $\binom{5}{3} + \binom{4}{2} + \binom{2}{1} = 18$. In general, the k-binomial expansion of a number provides a direct way of writing down the mth sequence in S_k.

k-sequences in S_k can be associated with k-sets of positive integers in the obvious way, and we will regard these two notions as freely interchangeable. In the lexicographic ordering of S_k the initial segment of length m forms a "cascade" of sets (Kruskal [45]) which can be described as follows. If

$$m = \binom{a_k}{k} + \binom{a_{k-1}}{k-1} + \cdots + \binom{a_1}{1},$$

then the first m sets in S_k consist of:

all k-sets in $[0, a_k - 1]$,

all k-sets formed by adding a_k to a $(k-1)$-set in $[0, a_{k-1} - 1]$,

all k-sets formed by adding $\{a_k, a_{k-1}\}$ to a $(k-2)$-set in $[0, a_{k-2} - 1]$,

.
.
.

and so forth.

The next lemma explains the operation of "lowering denominators" in a k-binomial expansion.

LEMMA 8.3: *Let $F = \{\sigma_0, \sigma_1, \ldots, \sigma_{m-1}\}$ be the initial segment of length m in S_k, and let ∂F denote the collection of $(k-1)$-sets which are subsets of members of F. If*

$$m = |F| = \binom{a_k}{k} + \binom{a_{k-1}}{k-1} + \cdots + \binom{a_i}{i},$$

then

$$|\partial F| = \binom{a_k}{k-1} + \binom{a_{k-1}}{k-2} + \cdots + \binom{a_i}{i-1}.$$

The proof is an immediate consequence of Lemma 8.2. We can thus restate the Kruskal-Katona Theorem as follows:

For $F \subseteq S_k$, $|F| = m$, the cardinality of ∂F is minimized by taking F to be the first m sets in S_k, in lexicographic order.

Another useful reformulation of Theorem 8.1 is based on notation introduced by Clements and Lindstrom [7]: given a family F of k-sets, define the *compression of F* (denoted CF) to be the family consisting of the first $|F|$ sets in S_k. We say that a family of F is *compressed* if $CF = F$.

It is trivial to verify that if F is compressed then ∂F is also compressed. Using Lemma 8.3, we can interpret the Kruskal-Katona Theorem as a statement about how the operators ∂ and C commute: for any family F of k-sets,

$$\partial CF \subseteq C\partial F.$$

Equivalently, if $F \subseteq S_k$ and $G \subseteq S_{k-1}$ and $\partial F \subseteq G$, then $\partial CF \subseteq CG$.

If F is a family of sets of varying size, we can interpret the operators C and ∂ as acting on each rank of F separately. A family K is called a *simplicial complex* (or *order ideal of sets*) if $\partial K \subseteq K$. By the above remarks, the following is also equivalent to Theorem 8.1:

THEOREM 8.4: *If K is a finite simplicial complex then so is its compression CK.*

Hence we can regard CK as a *canonical form* for simplicial complexes having a specified number of faces of each size.

If K is any simplicial complex, we define the *f-sequence* of K to be the sequence $\vec{f}(K) = (f_0, f_1, f_2, \ldots,)$ where f_i denotes the number of i-sets (or $(i - 1)$-faces) in K. Theorem 8.1 permits a complete characterization of those sequences of integers which arise as f-sequences. It is convenient to introduce the following notation: if

$$m = \binom{a_k}{k} + \binom{a_{k-1}}{k-1} + \cdots + \binom{a_i}{i},$$

$$a_k > a_{k-1} > \cdots > a_i \geq i > 0,$$

define

$$\partial_k(m) = \binom{a_k}{k-1} + \binom{a_{k-1}}{k-2} + \cdots + \binom{a_i}{i-1}.$$

THEOREM 8.5: *A sequence of integers $\vec{f} = (f_0, f_1, f_2, \ldots)$ is the f-sequence of some finite simplicial complex if and only if for each $k \geq 1$, $\partial_k(f_k) \leq f_{k-1}$.*

The proof is immediate: to construct a complex K with $\vec{f}(K) = (f_0, f_1, f_2, \ldots)$ take the first f_i sets of size i for each i. By Lemma 8.3, these sets form a simplicial complex.

By similar reasoning, it is possible to obtain canonical forms for *antichains* having a fixed number of sets of each size. If F is an antichain in B_n, let f_i again denote the number of i-sets in F, and define $\vec{f}(F) = (f_0, f_1, f_2, \ldots)$. The following theorem is due to Clements [6] and independently to Daykin, Godfrey, and Hilton [12]:

THEOREM 8.6: *Let $\vec{f} = (f_0, f_1, f_2, \ldots, f_n)$ be a sequence of nonnegative integers, and let k and l be the smallest and largest indices i for which $f_i \neq 0$. Then $\vec{f} = \vec{f}(F)$ for some antichain $F \subseteq B_n$ if and only if*

(***)
$$f_k + \partial_{k+1}(f_{k+1} + \partial_{k+2}(f_{k+2} + \cdots + \partial_{l-1}(f_{l-1} + \partial_l(f_l))\cdots) \leq \binom{n}{k}.$$

Theorem 8.6 essentially states that $\vec{f} = \vec{f}(F)$ for some antichain $F \subseteq B_n$ if and only if it is possible to construct F in the following canonical way: take the first f_l l-sets in lexicographic order; then take the next f_{l-1} $(l-1)$-sets which are available (i.e., which are not subsets of sets already chosen); then the next f_{l-2} $(l-2)$-sets, and so forth.

To prove Theorem 8.6, observe that for any antichain F, $\vec{f}(F)$ must satisfy (***) since the left hand side represents the smallest possible number of k-sets which are contained in members of F. Conversely, if $\vec{f}$ satisfies (***), it is easy to see that the canonical construction described above can always be carried out.

By analogy with simplicial complexes, we define the *compression CF* of an antichain F to be the canonical antichain whose f-sequence is $\vec{f}(F)$.

Every antichain F determines a unique simplicial complex (denoted by $K(F)$) whose maximal elements are the members of F. ($K(F)$ is sometimes called the *order ideal generated by F*.) It follows immediately from the definitions that F is compressed (as an antichain) if and only if $K(F)$ is compressed (as a simplicial complex). Equivalently, F is compressed if and only if $F = K - \partial K$ for some compressed simplicial complex K.

We can restate these observations in the form of a slight generalization of the Kruskal-Katona Theorem:

COROLLARY 8.7: *Let F be an antichain of sets in B_n, with $\vec{f}(F) = (f_0, f_1, f_2, \ldots)$. Let k be the smallest index for which $f_k \neq 0$, and let $l \leq k$. Then F dominates the smallest possible number of l-sets (among all antichains in B_n with the same f-sequence) if F is compressed.*

The inequality (***) in Theorem 8.6 can be viewed as a refinement of the fundamental LYM inequality (Section 4), if we divide both sides by $\binom{n}{k}$. For example, if F consists only of sets of size k and $k+1$, then (***) becomes

$$\frac{f_k}{\binom{n}{k}} + \frac{\partial_{k+1}(f_{k+1})}{\binom{n}{k}} \leq 1.$$

This is stronger than the LYM inequality since

$$\frac{\partial_{k+1}(f_{k+1})}{\binom{n}{k}} \geq \frac{f_{k+1}}{\binom{n}{k+1}},$$

as can be seen from Lemma 4.5 (the normalized matching property for sets).

Daykin [9] observed that the Kruskal-Katona Theorem can be used to give a short proof of the Erdös-Ko-Rado Theorem. We will discuss this argument next, as well as a number of extensions and refinements. In fact, we will prove a more general result, due to Kleitman [42], from which the Erdös-Ko-Rado Theorem follows immediately.

THEOREM 8.8: *Let F be a family of k-subsets of $\{1, 2, \ldots, n\}$, and let G be a family of l-subsets. Suppose that $k + l \leq n$, and that no member of F is disjoint from a member of G. If $|F| \geq \binom{n-1}{k-1}$ then $|G| \leq \binom{n-1}{l-1}$.*

To prove Theorem 8.8, let $\overline{F}$ denote the family of $(n - k)$-sets which are complements of members of F. Then $|\overline{F}| = |F|$, and the conditions on F and G imply that no member of $\overline{F}$ contains a member of G—that is, $\overline{F} \cup G$ is an antichain. By Theorem 8.6,

$$|G| + \partial_{l+1} (\partial_{l+2} (\cdots \partial_{n-k} (|F|) \cdots)) \leq \binom{n}{l},$$

which implies $|G| \leq \binom{n}{l} - \binom{n-1}{l} = \binom{n-1}{l-1}$, since $|F| \geq \binom{n-1}{n-k}$.

We have actually proved more:

COROLLARY 8.9: *If F and G are as in the statement of Theorem 8.8, except that $|F|$ is arbitrary, then*

$$|G| \leq \binom{n}{l} - \partial_{l+1} (\partial_{l+2} (\cdots \partial_{n-k} (|F|) \cdots)).$$

As a consequence of the proof of Theorem 8.8 we obtain the following corollary: If F is any collection of k-sets in B_n, let C^*F denote the *reverse compression of F in B_n*. That is, C^*F consists of the *last* $|F|$ sets in the lexicographic ordering of k-subsets of $\{1, 2, \ldots, n\}$. (Equivalently, $C^*F = \overline{CF}$, where a bar denotes taking all sets which are complements of sets in the given family.)

COROLLARY 8.10: *If F and G are families of k-sets and l-sets which satisfy the conditions of Theorem 8.8, then C^*F and C^*G also satisfy these conditions.*

Thus the pairs C^*F, C^*G of families which are "reverse-compressed" form a set of *canonical forms* for pairs of families satisfying the conditions of Theorem 8.8. Not every pair of families which are reverse-compressed has the "pairwise nondisjointness property," however. For this is to be true, it is necessary that the inequality in Corollary 8.9 hold.

When $k = l$ and $F = G$, Theorem 8.8 reduces to the Erdös-Ko-Rado Theorem, and Corollary 8.10 can be restated as follows:

COROLLARY 8.11: *Let F be a family of k-sets in B_n, $k \leq n/2$, which satisfies the conditions of the Erdös-Ko-Rado Theorem. Then C^*F also satisfies these conditions. In fact, this is true because every member of C^*F contains the element n.*

Using the above arguments, it is possible to obtain considerable refinements of the Erdös-Ko-Rado Theorem for arbitrary antichains (i.e., when the condition of uniform size is removed). In fact, the next theorem completely characterizes the "f-sequences" of antichains with the Erdös-Ko-Rado property, improving an earlier partial result of the "LYM" type (Theorem 5.4).

THEOREM 8.12: *Let $\vec{f} = (f_0, f_1, f_2, \ldots)$ with $f_0 = 0$ and $f_i \neq 0$ only if $i \leq n/2$. Then there exists an antichain $F \subseteq B_n$ whose members are pairwise nondisjoint, with $\vec{f}(F) = \vec{f}$, if and only if there exists an antichain $F' \subseteq B_{n-1}$ with $\vec{f}(F') = (f_1, f_2, f_3, \ldots)$.*

In other words, the f_i's corresponding to Erdös-Ko-Rado families on a set of size n behave exactly like those of Sperner families on a set of size $n - 1$. Necessary and sufficient conditions for sequences of the latter type are provided by Corollary 8.6.

To prove Theorem 8.12, we argue as follows: suppose that $F \subseteq B_n$ is an antichain which satisfies the given conditions. Let $C^*F = \overline{CF}$. Intuitively, C^*F is obtained by "reverse-compressing" each rank of F, beginning with the smallest ranks first. Clearly C^*F is an antichain with $\vec{f}(F) = \vec{f}(C^*F)$. The proof will be complete if we can show that each member of C^*F contains the element n, for then the required $F' \subseteq B_{n-1}$ can be obtained from C^*F by removing n from each set. Let l denote the largest index for which $f_l \neq 0$, and let $[F]^l$ denote the result of "projecting F up to rank l", i.e., $[F]^l$ consists of all l-sets which contain members of F. Similarly, define $[C^*F]^l$ to be the projection of C^*F up to rank l. Clearly $[F]^l$ is an Erdös-Ko-Rado family, since F itself is. Moreover, by Corollary 8.11 the reverse compression of $[F]^l$ has the property that each member contains n. But it is immediate from the definition of C^*F that $[C^*F]^l$ is reverse-compressed. Moreover $|[F]^l| \geq |[C^*F]^l|$ by Corollary 8.8 (applied upside down), and hence each member of $[C^*F]^l$ contains n. It follows easily that each member of C^*F contains n, and the proof is complete.

Reverse-compressed antichains, all of whose members contain the element n, can be thought of as canonical forms for Erdös-Ko-Rado families having a specified f-sequence.

Next we consider extensions of the Kruskal-Katona Theorem to lattices of multisets $M_{\vec{e}}$, when $\vec{e} = (e_0, e_1, \ldots)$, $e_i \in Z^+ \cup \{\infty\}$. We extend the notation introduced earlier: if $F \subseteq M_{\vec{e}}$, let ∂F denote the family of multisets of the form $(\sigma_0, \sigma_1, \ldots, \sigma_i - 1, \ldots)$, where $\sigma = (\sigma_0, \sigma_1, \ldots, \sigma_i, \ldots)$ is a member of F. Let S_k denote the set of elements of rank k in $M_{\vec{e}}$. Again order the elements of S_k in reverse lexicographic order (i.e., by the rightmost position in which they differ), and define the *compression* of a set $F \subseteq S_k$ (denoted CF) to be the initial segment of length $|F|$ in S_k.

Clements and Lindstrom [7] proved that for any family $F \subseteq S_k$, the identity

$$\partial CF \subseteq C\partial F$$

holds, provided that $e_0 \geq e_1 \geq e_2 \geq \cdots$. This identity leads to an analog of the Kruskal-Katona Theorem for lattices of multisets, which we restate as follows:

THEOREM 8.13: *Let F denote a family of multisets of rank k in $M_{\vec{e}}$, where $\vec{e} = (e_0, e_1, \ldots)$ and $e_0 \geq e_1 \geq e_2 \geq \cdots$. If F is fixed, then $|\partial F|$ is minimized by taking F to be compressed.*

Theorem 8.13 can be used to characterize the f-vectors of generalized "simplicial complexes" in lattices of the form $M_{\vec{e}}$. The characterization is analogous to that already obtained for sets (Theorem 8.5). The case $\vec{e} = (\infty, \infty, \infty, \ldots)$ of Theorem 8.13 is due to Macaulay [49].* (Another proof was given by Sperner [56].)

The lower bound on $|\partial F|$ can also be expressed in numerical form, using the notation introduced in section 1: one can show that $|F|$ has a unique decomposition

$$|F| = \binom{e_0, \ldots, e_{a_k}}{k} + \binom{e_0, \ldots, e_{a_{k-1}}}{k-1} + \cdots + \binom{e_0, \ldots, e_{a_i}}{i},$$

($i > 0$) where each term is nonzero, and $a_k \geq a_{k-1} \geq \cdots \geq a_i$ with no integer j repeated more than e_j times. The conclusion of Theorem 8.13 is that

$$|\partial F| \geq \binom{e_0, \ldots, e_{a_k}}{k-1} + \binom{e_0, \ldots, e_{a_{k-1}}}{k-2} + \cdots + \binom{e_0, \ldots, e_{a_i}}{i-1}.$$

If $\vec{e} = (1, 1, 1, \ldots)$, then $a_k > a_{k-1} > \cdots > a_i$, and these expressions reduce to the standard binomial decompositions. If $\vec{e} = (\infty, \infty, \infty, \ldots)$, there is no restriction on the a's, and we get a "negative binomial decomposition":

---

* Macaulay's object in studying this question was to obtain a characterization of *Hilbert functions* of certain kinds of modules. He considered R-modules of the form R/I, where polynomial ring in finitely many variables over a field, an I is a homogeneous ideal in R. A function $H: Z^+ \to Z^+$ is the Hilbert function of such a module if and only if there exists a generalized simplicial complex $F \subseteq M_{\vec{e}}$, where $\vec{e} = (\infty, \infty, \infty, \ldots)$ such that $\vec{f}(F) = (H(0), H(1), H(2), \ldots)$.

$$|F| = \left|\binom{-a_k}{k}\right| + \left|\binom{-a_{k-1}}{k-1}\right| + \cdots + \left|\binom{-a_i}{i}\right|$$

$$= \binom{a_k + k - 1}{k} + \binom{a_{k-1} + k - 2}{k - 1} + \cdots + \binom{a_i + i - 1}{i}$$

with $a_k \geq a_{k-1} \geq \cdots \geq a_i > 0$.

Many of the results mentioned earlier for lattices of sets can be extended to lattices of multisets using Theorem 8.13. We mention one example, which extends the Erdös-Ko-Rado Theorem in a direction which is not obvious at first glance. Before stating the general result, we begin with a special case, which is most conveniently stated in integer-divisor form. Recall that the "rank" of an integer is the total number of prime factors which appear in it.

THEOREM 8.14: *Let* $N = \prod_{i=1}^{m} p_i{}^{e_i}$, *where* $e_0 \geq e_1 \geq \cdots \geq e_m$ *and* e_m *is odd. Let F be an antichain of divisors of N, each having rank k (with $k \leq \Sigma e_i/2$). Suppose that for all a, b, $\in$ F, ab does not divide N. Then*

$$|F| \leq \binom{e_0, e_1, \ldots, e_m - \alpha}{k - \alpha}$$

where $\alpha = [e_m/2] + 1$.

Theorem 8.14 states that $|F|$ is maximized by taking all rank k divisors of n which have p_m^α as a factor, provided that e_m is odd. If n is square free (i.e., $e_0 = e_1 = \cdots = e_m = 1$) this result is equivalent to the Erdös-Ko-Rado Theorem since $ab \mid n$ if and only if the sets of primes occurring in a and b are disjoint.

If e_m is even, the statement must be modified as follows:

THEOREM 8.15: *Let N and F be as in Theorem 8.14, except that* e_m *is arbitrary. Let* G_0 *denote the set of divisors of N of the form* $p_0{}^{f_0} p_1{}^{f_1} \cdots p_m{}^{f_m}$, *where* $f_i > e_i/2$ *for some index i and* $f_j = e_j/2$ *for* all $j > i$. Let F_0 denote the elements of G_0 which have rank k. Then $|F| \leq |F_0|$.

It is easy to verify that $ab \nmid N$ for all $a, b \in G_0$, and hence F_0 is a family of the desired type. If e_m is odd, then G_0 consists of all $a|N$ such that $p^\alpha|a$, and Theorem 8.14 follows as a special case.

To prove Theorem 8.15, let $\overline{F}$ be the set of all divisors of N of the form N/a, $a \in F$, and define $\overline{F_0}$ similarly. Let $\overline{F}^*$ and $\overline{F_0}^*$ denote the "descendants" of $\overline{F}$ and $\overline{F_0}$ at level k. Clearly both $F \cup \overline{F}$ and $F_0 \cup \overline{F_0}$ are antichains. Moreover it is trivial to check that $\overline{F_0}^* = S_k - F_0$. Note that $\overline{F_0}$ is an "initial segment" in the reverse lexicographic ordering of $S_{\Sigma e_i - k}$, and hence so is $\overline{F_0}^*$. Hence by Theorem 8.13, if $|F| > |F_0|$, then $|\overline{F}^*| \geq |\overline{F_0}^*| = |S_k - F_0| > |S_k - F|$, which is impossible since $\overline{F}^* \subseteq S_k - F$.

In the statement of theorem 8.15, it is not necessary to assume that all of the members of F have rank k. The same result holds if F is any antichain whose members all have rank $\leq k$.

If we remove the conditions that F be an antichain from the hypotheses of Theorem 8.15, and also the restriction on ranks, we obtain the following:

THEOREM 8.16: *Let* $N = \prod_{i=0}^{m} p_i^{e_i}$, *with* $e_0 \geq e_{1'} \geq \cdots \geq e_m$. *Let* G *be an arbitrary set of divisors of* N, *such that* $ab \nmid N$ *for all* $a, b \in G$. *Then* $|G| \leq |G_0|$, *where* G_0 *is as defined in Theorem 8.15.*

The proof is immediate: for each $a \in G_0$, at most one of $\{a, N/a\}$ can be in G, and every divisor is either a or N/a for some $a \in G_0$. Trivially,

$$|G_0| = \begin{cases} \dfrac{1}{2} |M_{\vec{e}}|, & \text{if } N \text{ is not a square,} \\ \dfrac{1}{2} (|M_{\vec{e}}| - 1), & \text{if } N \text{ is a square.} \end{cases}$$

It is interesting to consider what happens when, in the lattice of divisors of an integer, the condition $ab \nmid N$ is replaced by the more natural condition $(a, b) = 1$. Erdös and Schönheim [18] obtained an analog of Theorem 8.16 for this case, but the analog of Theorem 8.15 seems more difficult, and little is known about it.

We conclude this section with a brief proof of Kruskal's theorem which is due to Clements and Lindstrom [7]. Their proof is actually more general, and applies to multisets as well (Theorem 8.13). We will give only the set-theoretic version, which is somewhat simpler:

Proof of Theorem 8.1: Our object is to prove that if $F \subseteq S_k$, $G \subseteq S_{k-1}$, and $\partial F \subseteq G$, then $\partial CF \subseteq CG$. Let n denote the largest index for which n is a member of some set in F or G. For each index, i, $0 \le i \le n$, define a new operator C_i (called *i-compression*) as follows: write $S_k(i) = \{A \in S_k \mid i \in A\}$ and $\overline{S_k(i)} = \{A \in S_k \mid i \notin A\}$. Then both $S_k(i)$ and $\overline{S_k(i)}$ inherit a linear ordering from the lexicographic ordering defined on S_k. For any family F of k-sets, define the i-compression of F (denoted C_iF) to be the union of the first $|F \cap S_k(i)|$ members of $S_k(i)$ and the first $|F \cap \overline{S_k(i)}|$ members of $\overline{S_k(i)}$. Clearly $|C_iF| = |F|$.

The proof of Theorem 8.1 is by induction on n, and is based on four elementary observations.

(1) If $F \subseteq S_k$, $G \subseteq S_{k-1}$, and $\partial F \subseteq G$, then $\partial C_iF \subseteq C_iG$, for all $i \le n$. (The i-compression of a simplicial complex is again a simplicial complex.)

(2) After repeated application of C_i for various i, F and G are transformed into sets F' and G' (with $|F| = |F'|$ and $|G| = |G'|$) which are i-compressed for all $i \le n$.

(3) If $2k \ne n + 1$, then F' is compressed (that is, $F' = CF$) and the proof is complete.

(4) In the special case when $2k = n + 1$, it is possible for F' to be i-compressed for all i but not compressed. However, F' can be made compressed by exchanging its *last* member for its immediate predecessor, an operation which preserves the relation $\partial F' \subseteq G'$. Hence the theorem is true in every case.

Statement (1) follows from the inductive hypothesis. The argument is straightforward. Statement (2) is trivial. To see that (3) holds, let x be the last member of F' (in ordering of S_k), and let y be any element of S_k which precedes x. If $y \notin F'$, then x and y cannot agree in position i for any $i \leq n$, since F' is i-compressed. Hence x and y must be complements, which implies $n + 1 = 2k$. In this case, it is easy to see that $F' - x + y$ is compressed and satisfies $\partial(F' - x + y) \subseteq G'$, which proves statement (4), and completes the proof of Theorem 8.1.

Apparently nothing is known about analogs of the Kruskal-Katona Theorem for subspaces of a finite vector space. On the other hand several consequences of it are known to be valid (Hsieh's Theorem, the LYM inequality) so there is reason to hope that such a result might exist.

9. APPLICATION: THE LITTLEWOOD-OFFORD PROBLEM

We conclude by showing how some of the ideas, methods, and results presented earlier can be applied to a geometric problem concerning distributions of linear combinations of vectors. The question was first raised by Littlewood and Offord [47]:

Let $\vec{v}_1, \vec{v}_2, \ldots, \vec{v}_n$ be vectors in a Hilbert space V, each of magnitude at least one. What is the maximum number of linear combinations of the form

$$\sum_{i=1}^{n} \epsilon_i \vec{v}_i \ (\epsilon_i = 0 \text{ or } 1)$$

which can lie in a sphere of diameter 1?

The answer is $\binom{n}{\lceil \frac{n}{2} \rceil}$, independently of the dimension of V, and this number can be achieved by taking all of the vectors to be the same. The solution to the Littlewood-Offord problem was conjectured by Erdös [15] and proved in several stages by Erdös, Katona, and Kleitman. Since the earlier arguments are elegant and elementary, we have included them below, together with the final dimension-free proof due to Kleitman.

Proof (when dim $V = 1$, (Erdös [15])): We can assume that all of the vectors are positive, since changing the sign of a vector only translates the set of linear combinations. Now to each linear combination associate the set of indices for which $\epsilon_i = 1$. Clearly, if a collection of linear combinations lies in a unit interval, the corresponding sets must form an antichain. Hence, by Sperner's Theorem, the number of linear combinations can be at most

$$\binom{n}{\left[\frac{n}{2}\right]}.$$

Proof (when dim $V = 2$, (Katona [31], Kleitman [35])): As above, we can change the direction of vectors if necessary and assume that the vectors all lie in two quadrants (say the first and second). Now we associate two sets of indices to each linear combination—one corresponding to vectors in the first quadrant and the other corresponding to vectors in the second quadrant. Since the sum of two unit vectors in a quadrant has magnitude at least $\sqrt{2}$ and lies in the same quadrant, we can deduce that two linear combinations lying within a unit diameter circle cannot have sets of indices which agree in one quadrant and are comparable in the other. The conditions on the pairs of index sets are therefore precisely those of Theorem 4.5, from which the bound of $\binom{n}{\left[\frac{n}{2}\right]}$ follows as before.

Proof (when dim V is arbitrary, (Kleitman [39])): The idea of this proof is to construct "saturated partitions" for linear combinations of vectors, imitating the methods of section 3. We will show that the collection of all linear combinations can be partitioned into $\binom{n}{\left[\frac{n}{2}\right]}$ blocks, such that no two linear combinations in a block can lie in the same unit diameter sphere. Trivially, this implies a bound of $\binom{n}{\left[\frac{n}{2}\right]}$.

The construction is analogous to the inductive procedure of de-Bruijn, Tengbergen, and Kruyswijk, in which a k-chain in B_{n-1} produces two new chains in B_n, one of length $k + 1$ and the other of length $k + 1$. We proceed in exactly the same way: suppose that the linear combinations of $\vec{v}_1, \vec{v}_2, \ldots, \vec{v}_{n-1}$ have been appropriately partitioned into blocks. Consider one such block U of size k, whose members we denote by $\vec{\gamma}_1, \vec{\gamma}_2, \ldots, \vec{\gamma}_k$. We can obtain two new blocks from U by taking all linear combinations with and without $\vec{v}_n$, denoting these blocks by $U + \vec{v}_n$ and U, respectively. Next take that linear combination $\vec{\gamma}_i \in U$ which has maximal component in the direction of $\vec{v}_n$, and transfer $\vec{\gamma}_i + \vec{v}_n$ from $U + \vec{v}_n$ to U. This gives new blocks of size $k - 1$ and $k + 1$, and one can easily check that no two members of either block lie in a sphere of diameter one. Repeating this construction for each block gives a partition of all linear combinations of $\vec{v}_1, \vec{v}_2, \ldots, \vec{v}_n$. Since the number of blocks of each size propagates in the same way as the number of chains in a partition of B_n, there must be $\binom{n}{\left[\frac{n}{2}\right]}$ blocks altogether, and the proof is complete.

Final remark: Because of the large number of papers in the literature on this subject, and because some of the results described here have been rediscovered several times, there may have been cases where our attribution of results has been incomplete. We apologize in advance for any such occurrences, and hope that the authors involved will inform us of any errors in the references. Also, requirements of space have forced us to leave unmentioned many interesting results which are closely related to the ones we have chosen to include. Again we offer apologies but suggest that the only remedy would be a much longer treatise on the subject.

The authors are grateful for helpful advice and comments from David E. Daykin, Richard P. Stanley, and David J. Kwiatkowsky.

REFERENCES

1. I. Anderson, "An application of a theorem of deBruijn, Tengbergen, and Kruyswijk", *J. Combinatorial Theory*, **3** (1967), 43–47.

2. K. Baker, "A generalization of Sperner's lemma", *J. Combinatorial Theory,* **6** (1969), 224-225.
3. B. Bollobas, "Sperner systems consisting of pairs of complementary subsets", *J. Combinatorial Theory* (A), **15** (1973), 363-366.
4. W. G. Brown, "Historical note on a recurrent combinatorial problem", *Amer. Math. Monthly,* 1965, 973-976.
5. N. deBruijn, C. A. van E. Tengbergen, and D. R. Kruyswijk, "On the set of divisors of a number", *Nieuw Arch. Wisk.,* (2) **23** (1952), 191-193.
6. G. Clements, "A minimization problem concerning subsets", *Discrete Math.,* **4** (1973), 123-128.
7. G. Clements and B. Lindstrom, "A generalization of a combinatorial theorem of Macaulay", *J. Combinatorial Theory,* **7** (1969), 230-238.
8. H. Crapo and G.-C. Rota, *Combinatorial Geometries,* M. I. T. Press, Cambridge, 1971.
9. D. E. Daykin, "Erdös-Ko-Rado from Kruskal-Katona", *J. Combinatorial Theory* (A), **17** (1974), 254-255.
10. ____, "A Simple Proof of Katona's Theorem", *J. Combinatorial Theory* (A), **17** (1974), 252-253.
11. D. E. Daykin and L. Lovasz, "The number of values of Boolean functions", *J. London Math. Soc.,* (2) **12** (1976), 225-230.
12. D. E. Daykin, J. Godfrey, and A. J. W. Hilton "Existence theorems for Sperner families", *J. Combinatorial Theory* (A), **17** (1974), 245-251.
13. R. P. Dilworth, "Some combinatorial problems on partially ordered sets", *Combinatorial Analysis,* R. Bellman, and M. Hall (eds.), *Proc. Symp. Appl. Math.,* Amer. Math. Soc., Providence, 1960, 85-90.
14. ____, "A decomposition theorem for partially ordered sets", *Ann. of Math.,* **51** (1950), 161-166.
15. P. Erdös, "On a lemma of Littlewood and Offord", *Bull. Amer. Math. Soc.,* **51** (1945), 898-902.
16. P. Erdös and D. Kleitman, "Extremal problems among subsets of a set", *Discrete Math.,* **8** (1974), 281-294.
17. P. Erdös, Chao Ko, and R. Rado, "Intersection theorems for systems of finite sets", *Quart. J. Math. Oxford Ser.* (2), **12** (1961), 313-318.
18. P. Erdös and J. Schönheim, "On the set of non-pairwise coprime divisors of a number", (to appear).
19. P. Erdös, M. Herzog, and J. Schönheim, "An extremal problem on the set of non-coprime divisors of a number", *Israel J. Math.,* **8** (1970), 408-412.
20. R. Freese, "An application of Dilworth's lattice of maximal antichains", *Discrete Math.,* **7** (1974), 107-109.
21. J. Goldman and G. C. Rota, "On the foundations of combinatorial theory IV: finite vector spaces and Eulerian generating functions", *Colloq. Math. Soc. Janos Bolyai*: 4. Combinatorial Theory and its Applications, *Balatonfured (Hungary),* 1969, 477-509.
22. R. Graham and L. H. Harper, "Some results on matching in bipartite graphs", *Studies in Applied Math., SIAM,* **4** (1970), 15-20.
23. C. Greene and D. J. Kleitman, "On the structure of Sperner k-families", *J. Combinatorial Theory* (A), **20** (1976), 41-68.

24. ____, "Strong versions of Sperner's theorem", *J. Combinatorial Theory* (A), **20** (1976), 80-88.
25. C. Greene, G. O. H. Katona, and D. J. Kleitman, "Extensions of the Erdös-Ko-Rado Theorem", *Studies in Applied Math.*, **55** (1976), 1-8.
26. P. Hall, "On representatives of subsets", *J. London Math. Soc.*, **10** (1935), 26-30.
27. L. H. Harper, "The morphology of partially ordered sets", *J. Combinatorial Theory*, **17** (1974), 44-58.
28. W. N. Hsieh, "Intersection theorems for systems of finite vector spaces", *Discrete Math.*, **12** (1975), 1-16.
29. W. N. Hsieh and D. J. Kleitman, "Normalized matching in direct products of partial orders", *Studies in Appl. Math.*, **52** (1973), 285-289.
30. G. O. H. Katona, "Extremal problems for hypergraphs", *Combinatorics*, M. Hall and J. H. vanLint (eds.), Mathematical Centre Tracts 56, Amsterdam, 1974.
31. ____, "On a conjecture of Erdös and a stronger form of Sperner's theorem", *Studia Sci. Math. Hungar.*, **1** (1966), 59-63.
32. ____, "Families of subsets having no subset containing another with small difference", *Nieuw. Arch. Wisk.*, (3) **20** (1972), 54-67.
33. ____, "A simple proof of the Erdös-Ko-Rado theorem", *J. Combinatorial Theory* (B), **13** (1972), 183-184.
34. ____, "A theorem of finite sets", *Proc. Tihany Conf.*, 1966, Budapest, 1968.
35. D. J. Kleitman, "On a lemma of Littlewood and Offord on the distribution of certain sums", *Math. Z.*, **90** (1965), 251-259.
36. ____, "On families of subsets of a finite set containing no two disjoint sets and their union", *J. Combinatorial Theory* (A), **5** (1968), 235-237.
37. ____, "On an extremal property of antichains in partial orders. The LYM property and some of its implications and applications", *Combinatorics*, M. Hall and J. H. vanLint, eds., Math. Centre Tracts 55, Amsterdam, 1974, 77-90.
38. ____, "Families of nondisjoint subsets'", *J. Combinatorial Theory*, **1** (1966), 153-155.
39. ____, "On a lemma of Littlewood and Offord on the distributions of linear combinations of vectors", *Advances in Math.*, **5** (1970), 1-3.
40. ____, "On a combinatorial problem of Erdös", *Proc. Amer. Math. Soc.*, **17** (1966), 139-141.
41. ____, "Extremal properties of collections of subsets containing no two sets and their union", *J. Combinatorial Theory* (A), **20** (1976), 390-392.
42. ____, "On a conjecture of Milner on k-graphs with non-disjoint edges", *J. Combinatorial Theory*, **5** (1968), 153-156.
43. D. J. Kleitman, M. Edelberg, and D. Lubell, "Maximal sized anti-chains in partial orders", *Discrete Math.*, **1** (1971), 47-53.
44. D. J. Kleitman and J. Spencer, "Families of k-independent sets", *Discrete Math.*, **6** (1973), 255-262.
45. J. Kruskal, "The number of simplices in a complex", *Mathematical Optimization Techniques*, University of California Press, Berkeley and Los Angeles, 1963, 251-278.

46. E. Levine and D. Lubell, "Sperner collections on sets of real variables", *Ann. N. Y. Acad. Sci.*, **75** (1970), 172-176.
47. J. E. Littlewood and C. Offord, "On the number of real roots of a random algebraic equation (III)", *Mat. USSR Sb.*, **12** (1943), 277-285.
48. D. Lubell, "A short proof of Sperner's theorem", *J. Combinatorial Theory*, **1** (1966), 299.
49. F. S. Macaulay, "Some properties of enumeration in the theory of modular systems", *Proc. London Math. Soc.*, **26** (1927), 531-55.
50. L. D. Meshalkin, "A generalization of Sperner's theorem on the number of subsets of a finite set", *Theor. Probability Appl.*, **8** (1963), 203-204.
51. G.-C. Rota, "A generalization of Sperner's theorem", *J. Combinatorial Theory*, **2** (1967), 104.
52. J. Schönheim, "A generalization of results of P. Erdös, G. Katona, and D. J. Kleitman concerning Sperner's theorem", *J. Combinatorial Theory* (A), **11** (1971), 111-117.
53. ____, "On a problem of Purdy", (to appear).
54. J. Spencer, "A generalized Rota conjecture for partitions", *Studies in Applied Math.*, **53** (1974), 239-241.
55. E. Sperner, "Ein Satz über Untermenge einer endlichen Menge", *Math Z.*, **27** (1928), 544-548.
56. ____, "Über einen kombinatorischen Satz von Macaulay und seine Anwendung auf die Theorie der Polynomideale", *Abh. Math. Sem. Univ. Hamburg*, **7** (1930), 149-163.
57. K. Yamamoto, "Logarithmic order of free distributive lattices", *J. Math. Soc. Japan*, **6** (1954), 343-353.
58. L. R. Ford and D. R. Fulkerson, *Flows in Networks*, Princeton University Press, Princeton, 1962.
59. D. J. Kleitman and E. C. Milner, "On the average size of sets in a Sperner family", *Discrete Math.*, **6** (1973), 141.
60. D. J. Kleitman, "On subsets contained in a family of non-commensurable subsets of a finite set", *J. Combinatorial Theory*, **1** (1966), 297-299.
61. G. Hansel, "Sur le Nombre des Fonctions Booléennes Monotones de n Variables", *C. R. Acad. Sci. Paris*, **262** (1966), 1088-1090.
62. D. J. Kleitman and G. Markowsky, "On Dedekind's problem: the number of monotone Boolean functions II", to appear in *Trans. Amer. Math. Soc.*
63. L. H. Harper and G.-C. Rota, "Matching Theory, an Introduction", *Advances in Probability and Related Topics* (P. Ney, ed.), Marcel Dekker, New York, 1971, 168-215.

RAMSEY THEORY

R. L. Graham and B. L. Rothschild

INTRODUCTION

In 1930, F. P. Ramsey [27] proved a remarkable theorem as part of his investigations in 'formal logic'. The theorem is a profound generalization of the 'pigeon hole principle' or 'Dirichlet box principle'. As is the case with many beautiful ideas in mathematics, Ramsey's Theorem extends just the right aspect of an elementary observation and derives consequences which are extremely natural although far from obvious. Recently it has been recognized that many results in combinatorial theory and other areas have the same flavor as Ramsey's Theorem, and the attempt to capture this common flavor, and to develop some general ideas based on it, has led to a proliferation of results which consitute what we describe here as 'Ramsey Theory'. In many cases, including the original theorem itself, the existence of certain numbers is asserted. A large effort has gone into finding exact values and bounds for these 'Ramsey numbers'.

We shall try here to describe some interesting aspects of the subject. We will of necessity leave out many more specialized results as well as all proofs, with the exception of several simple ones which we include at the end of the chapter. Rather we will try to emphasize the general ideas of Ramsey Theory, hopefully sacrificing completeness for the sake of palatability.

We now describe a certain very general class of theorems which we shall call Ramsey theorems. In contrast to many subjects, here the most general formulation is by far the most elementary and immediate.

A *bipartite graph* G is a graph with its vertices divided into two classes A and B, such that each edge of G has one vertex in A and one vertex in B. If the vertices of A are partitioned into r classes (called *colors*) then a vertex b in B is called *monochromatic* if all vertices a in A which are adjacent to b lie in a single class (where possibly r is infinite). G is said to have the Ramsey property for r colors, or to be *r-Ramsey*, if for every partition of A into r or fewer parts (called *r-colorings*), there is a monochromatic vertex in B (see Figure 1). The entire field of Ramsey Theory basically can be thought of as the attempt to decide which graphs are r-Ramsey.

Although this general formulation is very simple, and all Ramsey theorems can be expressed in this common form, more intuitively convenient formulations are used for individual cases. Below we give a few of the more appealing examples of Ramsey theorems.

SOME EXAMPLES

One of the earliest of the Ramsey theorems is the following theorem of Schur [28]:

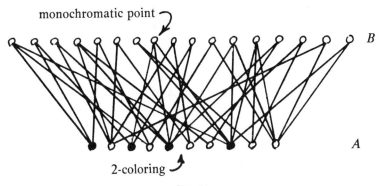

Fig. 1.

Example 1. For every positive integer r, there is an $N = N(r)$ such that if the set $[1, N]$ (integers x, with $1 \leq x \leq N$) is r-colored, then there must exist $x, y, z \in [1, N]$ all having the same color and satisfying $x + y = z$.

In terms of bipartite graphs, we define a graph G as follows. Let $A = [1, N]$, $B = \{\{x, y, z\} : x + y = z, x, y, z \in [1, N]\}$, and for $a \in A$, $b \in B$, let $\{a, b\}$ be an edge of G iff $a \in b$. That the bipartite graph G is r-Ramsey is Schur's theorem.

An example of a result not usually thought of as a Ramsey theorem is the Baire Category Theorem.

Example 2. Let A be the set of points of a complete metric space X. Let B be the set of subsets $S \subseteq X$ such that the closure $\overline{S}$ contains an open set of X. For $a \in A$, $b \in B$, let $\{a, b\}$ be an edge iff $a \in b$. Then this graph is $\aleph_0$-Ramsey.

This is just the statement that a complete metric space is not the countable union of nowhere dense sets. Of course, even though this theorem is of the Ramsey form, it is not really a combinatorial theorem.

The next example is Ramsey's original theorem.

Example 3. Let A be the set of all k element subsets of a countable set S. Let B be the set of all infinite subsets of S. For $a \in A$, $b \in B$, let $\{a, b\}$ be an edge iff $a \subseteq b$. Then this graph is r-Ramsey for all positive integers r.

Ramsey also proved a finite version of this theorem (where B is the set of all l element subsets of a sufficiently large finite set S') which we treat in the next section. We remark here that the reader is probably already familiar with the simplest case of this theorem, namely, that if all the edges of the complete graph on 6 vertices are 2-colored, then some monochromatic triangle must be formed. To see this, simply observe that for any vertex v there must be three vertices v_1, v_2, v_3 with all edges $\{v, v_i\}$ having the same color. If any edge $\{v_i, v_j\}$ also had this color, we are done. On the other hand, if all the edges $\{v_i, v_j\}$ have the opposite color, we are also done.

Example 4. Let A be the set of all subsets of cardinality 2 of a set S of cardinality $2^{\aleph_0 +}$. (If α is a cardinal, α^+ denotes the next largest cardinal.) Let B be the set of all subsets of S of cardinality $2^{\aleph_0}$. For $a \in A$, $b \in B$, let $\{a, b\}$ be an edge iff $a \subseteq b$. Then this graph is 2-Ramsey.

This is a theorem of Erdös. Replacing $2^{\aleph_0 +}$ by $2^{\aleph_0}$, Dushnik and Miller [7] show that the resulting graph is not 2-Ramsey. These results belong to a vast and still growing literature on Ramsey theorems for large cardinals, ordinals and order types. The assertions that certain graphs of this sort are Ramsey sometimes turn out to be independent of the usual axioms for set theory [10].

Example 5 (van der Waerden [30]). Let r and k be positive integers. Let $A = [1, W]$ and let B be the set of all k term arithmetic progressions in $[1, W]$. For $a \in A$, $b \in B$, let $\{a, b\}$ be an edge iff $a \in b$. Then there is a function $W(k, r)$ such that if $W \geq W(k, r)$ then the graph is r-Ramsey.

This theorem asserts that if we r-color a sufficiently large interval of integers, it must contain a monochromatic arithmetic progression of k terms. In fact, an awesome result has recently been proved by E. Szemerédi [29], settling a 40 year old conjecture of Erdös and Turán. Namely:

If R is a subset of the positive integers with positive upper density, i.e.,

$$\lim_N \sup \frac{|R \cap [1, N]|}{N} > 0,$$

then R contains arbitrarily long arithmetic progressions.

This theorem is equivalent to a deceptively simple strengthening of van der Waerden's Theorem. It is the statement that for Example 5, not only is there a monochromatic progression of length k, but in fact there must be such a progression having that color which occurs most frequently in $[1, W]$.

CATEGORIES

Until now we have considered individual graphs and their Ramsey properties. However, many of the most important theorems and their proofs, including Ramsey's Theorem itself, require the consideration of whole families of graphs. K. Leeb [21] pointed out that in such situations the use of category theory can be quite helpful both in the formulation and in the proofs of results. The introduction of categorical methods was initially inspired by the attempt to prove the insightful conjecture of G.-C. Rota, namely, that the analogue to Ramsey's Theorem for finite vector spaces holds. It is now known that certain categories are "Ramsey". These include the category of sets (Ramsey's Theorem) and the category of finite vector spaces (Rota's conjecture).

In their usual formulation these theorems are as stated below. The more formal statements in terms of categories will be given at the end of this section, where we state a general theorem for certain categories. These categories include the two above.

THEOREM [27]: *For all k, l, r, there exists a least integer $n(k, l, r)$ such that if $n \geq n(k, l, r)$ and the set of k-subsets of an n-set N is arbitrarily r-colored, all the k-subsets of some l-subset of N have a single color.*

THEOREM [14]: *For all k, l, r, there exists a least integer $n_q(n, l, r)$ such that if $n \geq n_q(n, l, r)$ and set of k-dimensional subspaces of an n-dimensional space N over $GF(q)$ is arbitrarily r-colored, then all the k-dimensional subspaces of some l-dimensional subspace have a single color.*

There are categories, however, for which the Ramsey property holds only in part. The proofs of the validity of the Ramsey property for the appropriate parts of these categories are among the nicest in the theory. For example, van der Waerden's theorem is one such case.

Consider the category P whose objects are the finite arithmetic progressions of the positive integers. Let the morphisms be the monomorphic affine maps from one progression to another (i.e., $x \to ax + b$). The Ramsey property for this category would assert: For every k, l, r, there is an $n = n(k, l, r)$ such that if the k term progressions in $[1, n]$ are r-colored, then all the k term subprogressions of some l term progression have the same color. For $k = 1$, this is just van der Waerden's theorem. However, for $k = 2$, it is false. To see this, consider the following 2-coloring of the two term arithmetic progressions:

$$c(\{x_1, x_2\}) = \alpha^* \text{ where } \alpha^* \equiv \alpha(\text{mod } 2), \alpha^* = 0 \text{ or } 1$$

and 2^α is the largest power of 2 dividing $x_2 - x_1$.

Another category for which the partial validity of the Ramsey property holds is the category G of finite, undirected graphs for which the morphisms are induced subgraph embeddings. φ is such an embedding if it is an injective mapping taking vertices of a finite graph L into vertices of a finite graph N and edges of L into edges of N such that for any two vertices u, v, of L, $\{u, v\}$ is an edge of L iff $\{\varphi(u), \varphi(v)\}$ is an edge of N. The Ramsey property for this category then says that for any graphs K, L and any integer r, there is a graph N such that for every r-coloring of the K-subgraphs of N (i.e., induced subgraphs isomorphic to K), there is an L-subgraph with all of its K-subgraphs having the same color. For the case in which K is a single vertex (and L is arbitrary), the Ramsey property holds by a result of Folkman [13]. In the case where K consists of a single edge, the Ramsey property also holds. This is a powerful new result of Deuber [6]. If K is a complete graph on k vertices and L is a complete graph on l vertices, then the Ramsey property becomes just the statement of Ramsey's Theorem.

Finally, if we consider the category G_n of graphs with no complete subgraph on n vertices (and the same morphisms as before), then the Ramsey property holds if K is a single vertex or if K is a single edge. The first result here is due to Folkman [13]. The second result is a major new result of Nešetřil and Rödl [23].

However, just as in the case of arithmetic progressions, easy counterexamples show that the Ramsey property does not hold in general in these categories. For example, in the category G, let K

be P_3, the tree on three vertices and let L be the 4-cycle C_4. Let N be an arbitrary graph on n vertices labelled with the elements of $[1, n]$. Color the tree $\underset{x}{\bullet}\!-\!\underset{y}{\bullet}\!-\!\underset{z}{\bullet}$ red if $y > \max(x, z)$ and blue otherwise. Clearly, no C_4 in N can have all four of its subgraphs P_3 with the same color. Similar counterexamples work for the categories G_n.

Very recently, Ramsey properties for certain categories of hypergraphs have been established in work of Nešetřil and Rödl [24]. Deeper understanding of these categories will be required before the full role of the Ramsey property is apparent.

We conclude this section with the Ramsey theorem for categories. The later sections will not depend on ideas from category theory.

Formally, we can define the Ramsey property for a category C as follows: If N and K are objects of C, then we let $C\begin{bmatrix} N \\ K \end{bmatrix}$ denote the set of subobjects of N of type K, where a subobject of N is said to be of *type* K if it contains a monomorphism $K \to N$. If $\varphi: L \to N$ is a monomorphism then we let $\bar\varphi : C\begin{bmatrix} L \\ K \end{bmatrix} \to C\begin{bmatrix} N \\ K \end{bmatrix}$ denote the obvious induced map.

The category C is called *Ramsey* if for every positive integer r and every pair of objects K and L, there is an object N such that for every r-coloring $c : C\begin{bmatrix} N \\ K \end{bmatrix} \to [1, r]$ there exists a monomorphism $\varphi: L \to N$ and an $i \in [1, r]$ such that the following diagram commutes:

$$\begin{array}{ccc} C\begin{bmatrix} N \\ K \end{bmatrix} & \xrightarrow{c} & [1, r] \\ {\scriptstyle\bar\varphi}\!\uparrow & & \uparrow{\scriptstyle\text{incl.}} \\ C\begin{bmatrix} L \\ K \end{bmatrix} & \longrightarrow & \{i\} \end{array}$$

For example, to state Ramsey's Theorem in this form, we consider the category S of finite sets with morphisms being injective mappings. We see that if N is an n-set, then its subobjects of type

K correspond to its subsets of $k = |K|$ elements. Note that $\left| S\begin{bmatrix} N \\ K \end{bmatrix} \right| = \binom{n}{k}$. Since all sets of the same cardinality are isomorphic, then the assertion that S is Ramsey is just Ramsey's original statement. For vector spaces, we consider the category $\mathcal{V}_q$ of finite-dimensional vector spaces over $GF(q)$ with morphisms being injective linear mappings. Here the subobjects of type K of an n-dimensional space N correspond to the $|K|$-dimensional subspaces of N.

THEOREM: *Let $\mathcal{C}$ be a class of categories such that for each category B in $\mathcal{C}$ there is a category A in $\mathcal{C}$ such that A and B satisfy the conditions below. Then all categories in $\mathcal{C}$ are Ramsey.*

The conditions are as follows, where we are assuming that all categories have as objects the set of nonnegative integers:

There is a function M from A to B with $M(l) = l + 1$, $l = 0, 1, 2, \ldots$, a functor P from B to A with $P(l) = l$, $l = 0, 1, 2, \ldots$, an integer $t \geq 0$, and for each $l \geq 0$, there are t monomorphisms $l \xrightarrow{\varphi_{lj}} l + 1, j \in [1, t]$, satisfying the following three conditions:

I. For each $k = 0, 1, 2, \ldots$, the diagonal d in the following diagram is epic, where $\amalg$ (together with the indicated injections) denotes coproduct, and d is the unique map determined by the coproduct which makes the diagram commute:

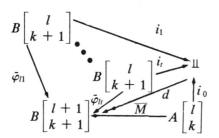

Here, $\bar{M}$ is the mapping induced on subobjects by M.

II. For each $s \xrightarrow{g} l$ in B and each $j \in [1, t]$, the following diagram commutes:

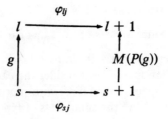

III. For some $l \xrightarrow{e} l + 1$ in A, the following diagram commutes for all $j \in [1, t]$:

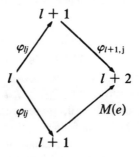

For $\mathcal{C}$ consisting of the single category $\mathcal{S}$ with morphisms $k \to l$ being the monomorphisms from $[1, k]$ to $[1, l]$, the theorem (with appropriate choices M, P and $t = 1$) becomes Ramsey's Theorem. In this case only condition I remains interesting, reflecting simply the basic relation for binomial coefficients, namely, $\binom{l+1}{k+1} = \binom{l}{k} + \binom{l}{k+1}$. In particular, the $(k + 1)$-subsets of an $(l + 1)$-set consist of those which contain a fixed element x together with those not containing x. The relation above is an example of a much more general "Pascal relation", as described by Leeb [22]. The study of such general theorems has sometimes been referred to as 'Pascal Theory'. Other categories, including those of finite binary trees (with appropriate inclusion morphisms) and finite

Boolean algebras (with sublattice monomorphisms) also have Pascal relations and the Ramsey property.

RAMSEY NUMBERS

Most of the finite Ramsey theorems we have considered up to this point involve the existence of "Ramsey numbers". In this section we will discuss some of the known values and bounds for these numbers.

We recall that Ramsey's Theorem asserts the existence of a least integer $n(k, l, r)$ such that any r-coloring of the k-subsets of an n-set S forces all the k-subsets of some l-subset of S to have a single color, provided $n \geq n(k, l, r)$. All the values of $n(k, l, r)$ currently known are given by (e.g., see [18]):

$$n(1, l, r) = r(l - 1) + 1; n(l, l, r) = l;$$

$$n(2, 3, 2) = 6, n(2, 3, 3) = 17, n(2, 4, 2) = 18.$$

The most thoroughly studied case has been $k = 2$, for which it is natural to phrase the results in terms of coloring the edges of a complete graph. We can define $R(l_1, \ldots, l_r)$ to be the least integer such that if $n \geq R(l_1, \ldots, l_r)$ and the edges of the complete graph K_n are arbitrarily r-colored, then there is a monochromatic K_{l_i} for some color i. The numbers known here [18] (which are not included above) are as follows:

$$R(3, 4) = 9, R(3, 5) = 14, R(3, 6) = 18, R(3, 7) = 23.$$

These are all the values known exactly. For other choices of the parameters, only estimates are available, some of which are:

$$27 \leq R(3, 8) \leq 30, 36 \leq R(3, 9) \leq 37,$$

$$c_1 k 2^{k/2} \leq R(k, k) \leq c_2 \frac{4^k}{\sqrt{k}} \frac{\log \log k}{\log k}, \qquad [9], [18]$$

$$R(x, y) \le c_3 y^{x-1} \frac{\log \log y}{\log y}, \qquad [18]$$

(a proof for the lower bound on $R(k, k)$ may be found in the paper by Joel Spencer in this volume).

For van der Waerden's theorem (Example 5), the estimates for the corresponding Ramsey numbers $W(k, r)$ are much less accurate. For small values, it is known that: $W(2, 2) = 3$, $W(3, 2) = 9$, $W(4, 2) = 35$ [4]. On the other hand, for larger values it is known that

$$k \, 2^k \le W(k, 2) < A(k, 4),$$

where the first inequality holds for k prime [1] and $A(m, n)$ is defined by:

$$A(1, n) = 2^n, A(m, 2) = 4, m \ge 1, n \ge 2,$$

$$A(m, n) = A(m - 1, A(m, n - 1)), m \ge 2, n \ge 3.$$

The reader is invited to calculate a few values of A, e.g., $A(5, 5)$ or $A(12, 3)$, in order to get a feeling for the disparity between these upper and lower bounds.

In general, the best constructive upper bounds for Ramsey numbers are strongly correlated to the complexity of the arguments used to show their existence. Hence, it is not surprising that the estimates for some of the more subtle Ramsey numbers are extremely large. For example, consider the Ramsey number $N(l)$ defined to be the least integer such that if $n \ge N(l)$ and the line segments between all pairs of vertices of a given n-dimensional rectangular parallelepiped P_n are arbitrarily 2-colored, then all the line segments between the pairs of vertices of some l-dimensional rectangular subparallelepiped of P_n have a common color (see [15] for a proof of existence). Of course, $N(1) = 1$. The best available estimate for $N(2)$, however, is:

$$6 \le N(2) \le A(A(A(A(A(A(A(12, 3), 3), 3), 3), 3), 3), 3).$$

It is conjectured that $N(2) = 6$.

SOME OLD DIRECTIONS

An important area we have yet to discuss originated some 40 years ago with the fundamental work of R. Rado [25], [26]. It deals with the integer solutions to systems of linear equations and contains as special cases both the theorem of Schur (Example 1) and van der Waerden's theorem.

Let $\mathcal{L} = \mathcal{L}(x_1, \ldots, x_n)$ denote a system of homogeneous linear equations in the variables $x_1, \ldots, x_n$ with integer coefficients. We say that $\mathcal{L}$ is *r-Ramsey* (called *r-regular* by Rado) if for any partition of the positive integers **P** into r classes, $\mathcal{L}$ has a solution $(a_1, \ldots, a_n)$ with all the a_k in one class. If $\mathcal{L}$ is r-Ramsey for all r then $\mathcal{L}$ is called *Ramsey*.

The basic result here is the following theorem of Rado [25]:

THEOREM: $\mathcal{L}$ *is Ramsey if and only if there is a partition of* $[1, n] = S_1 \cup \cdots \cup S_l$ *so that for each* $k \in [1, l]$, *there is a solution* $u^{(k)} = (u_1^{(k)}, \ldots, u_n^{(k)})$ *satisfying*

$$u_i^{(k)} = \begin{cases} 0, & \text{if } i \in S_j \text{ for } j > k, \\ 1, & \text{if } i \in S_k, \\ \text{arbitrary}, & \text{otherwise}. \end{cases}$$

For the special case in which $\mathcal{L}$ consists of a single equation, the result is particularly appealing:

COROLLARY: *The equation* $\sum_{k=1}^{m} a_k x_k = 0$ *is Ramsey if and only if some nonempty subset of the a_k's sum to zero.*

Not only does the above theorem imply van der Waerden's theorem and the previously mentioned theorem of Schur, but it also proves the following interesting generalization (see also [16]):

THEOREM: *Given integers k and r, there exists an integer $N(k, r)$ such that if $n \geq N(k, r)$, then for any r-coloring of $[1, n]$ there* exists $A \subseteq [1, n]$ with $|A| = k$ for which all the sums $\sum_{b \in B} b$, $\emptyset \neq B \subseteq A$, have the same color.

A striking extension of this result for the case of k infinite has recently been given by Hindman [20], [0].

THEOREM: *For any r and any r-coloring of the positive integers $\mathbf{P}$ there is an infinite set $A \subseteq \mathbf{P}$ such that all sums $\sum_{b \in B} b$, $\emptyset \neq B \subseteq A$, have the same color.*

An intriguing problem is to determine those infinite systems of homogeneous linear equations which are Ramsey.

We call a set $A \subseteq \mathbf{P}$ *regular* if any Ramsey system $\mathcal{L}$ of linear equations has a solution entirely in A. More than 35 years ago, Rado put forth the following conjecture: If a regular set of integers is partitioned into a finite number of classes then at least one of the classes is regular. This conjecture has very recently been proved in a study of Deuber [5] in which he characterizes regular sets of integers in terms of certain high-dimensional array-like subsets for which he is able to establish a Ramsey property.

SOME NEW DIRECTIONS

Certain questions have only recently been pursued. We discuss here two of the most active areas. The first of these involves the determination of numbers somewhat more general than those arising from Ramsey's Theorem, an immediate corollary of which is the following statement: For any choice of finite graphs $G_1, \ldots, G_r$, there is an $N = N(G_1, \ldots, G_r)$ such that if $n \geq N$ and the edges of K_n are r-colored, then there is an $i \in [1, r]$ and a subgraph (not necessarily induced) isomorphic to G_i with all its edges having color i. If $l = \max(|G_1|, \ldots, |G_r|)$ where $|G_i|$ denotes the number of vertices of G_i then clearly $N(G_1, \ldots, G_r) \leq n(2, l, r)$, the number from Ramsey's Theorem. The estimation of the "graph Ramsey numbers" provides some information on the size of the original

Ramsey numbers. However, in most of the cases considered up to now, the graphs G_i have been too simple for the resulting bounds on $N(G_1, \ldots, G_r)$ to be of much use in estimating Ramsey numbers. Some of these cases do turn out to have nice, exact answers and are themselves fascinating results in graph theory. For example,

$$N(P_m, P_n) = m + \left[\frac{n}{2}\right] - 1,$$

$$N(mK_3, nK_3) = 3m + 2n \text{ for } m \geq n, m \geq 2,$$

$$N(\underbrace{C_4, \ldots, C_4}_{r}) = r^2 + O(r),$$

where P_m denotes a path with m vertices, mK_3 denotes the union of m disjoint triangles and C_4 denotes a 4-cycle. For a survey of results in this direction, the reader is referred to Burr [2].

We have already seen one area of Ramsey theory where geometric considerations arise, namely, the Ramsey theorem for finite vector spaces. The theory is rather complete here, except for the calculation or estimation of the corresponding Ramsey numbers. When we consider Euclidean geometry, however, the corresponding theorem is clearly false. That is, there are 2-colorings of the points of Euclidean n-space $\mathbf{E}^n$ so that no Euclidean line is monochromatic. (For example, consider concentric spherical shells of alternating colors). The question of which monochromatic configurations must occur is the subject of "Euclidean Ramsey Theory" [11], [12]. In some sense, the obstruction to finding a monochromatic line is the fact that the underlying field is infinite, and consequently there are infinitely many points on each line. (Cates and Hindman [3] have investigated Ramsey properties for other infinite fields and for spaces of infinitely many dimensions, mostly along lines similar to the studies of the Ramsey properties of large cardinals.)

Thus, it is natural to consider finite configurations $C \subseteq \mathbf{E}^n$. Let G be a permutation group acting on $\mathbf{E}^n$. The problem is to determine those configurations C such that for any r-coloring of $\mathbf{E}^n$, for some $g \in G$, all the points of $g(C)$ have a single color.

For the case when G is the affine group on $\mathbf{E}^n$, it was shown by Gallai [26] that all finite configurations are Ramsey, i.e., are r-Ramsey for all r, thus generalizing van der Waerden's theorem. For the case when G is the identity group, then, of course, only one-point configurations are Ramsey.

An extremely interesting case is that of G being the group of Euclidean motions in $\mathbf{E}^n$. Thus, we must find a monochromatic set C' which is *congruent* to C. For the configuration C_1 consisting of two points separated by a fixed distance d, it is easily seen that any 2-coloring of $\mathbf{E}^2$ contains a monochromatic set congruent to C_1, e.g., by just considering the three vertices of an equilateral triangle of side d. A similar argument using higher dimensional simplexes shows that some regular simplex of side d will occur monochromatically in any r-coloring of a sufficiently high dimensional space. The minimum sufficient dimension $n(k, r)$ is a function of k, the number of vertices of the simplex, and r. At present, even the value of $n(r, 2)$ is unknown. It is known that $n(2, 2) = 2$, $n(7, 2) > 2$ and $n(2, 3) = 3$. In general, if for each r there is a monochromatic set congruent to C in any r-coloring of $\mathbf{E}^n$, provided only that n is sufficiently large depending upon r and C, then we say that C is Ramsey. At present the only configurations known [11] to be Ramsey are subsets of the vertices of a rectangular parallelepiped. In the other direction, it has been shown [11] that any Ramsey configuration must lie on some (perhaps very high dimensional) sphere. For configurations not in either of these classes, the simplest being the set of vertices of an obtuse triangle, it is not known whether any of them are Ramsey. The arguments for the spherical sets involve an extension of certain negative results of Rado (mentioned in the previous section) to a larger class of underlying fields than he considered.

One of the most appealing questions in this area is the conjecture that any 2-coloring of the Euclidean plane must contain a monochromatic set congruent to any given 3-set, with the possible exception of the set of vertices of a single equilateral triangle. Many families of triangles (i.e., 3-sets) are known to occur in this case, but for most triangles it is undecided [12]. Since this area is so new, very few of the obvious questions have been studied, e.g., other configurations, other groups G, and other geometries, to mention a few.

A FEW PROOFS

This section contains short proofs of the earliest Ramsey theorems, namely, Ramsey's Theorem, van der Waerden's theorem, Schur's theorem and a proof that the Ramsey number $r(K_3, K_3, K_3) = 17$ (due to Greenwood and Gleason [19]).

THEOREM (Ramsey): *For every k, l, r there is a least integer $n(k, l, r)$ such that if $n \geq n(k, l, r)$ and all the k-subsets of an n-set S are r-colored, then all the k-subsets of some l-subset of S have the same color.*

Proof: We use induction on k. For $k = 1$, this is just the box principle and $n(1, l, r) = (l - 1)r + 1$. Assume that the theorem holds for some $k \geq 1$ and all l and r. We prove the following: For each $j \leq k$, there is a number $f(j, k)$ such that if $n \geq f(j, k)$ and the $(k + 1)$-subsets of $[1, n]$ are r-colored, then there exists a k-subset $H = X \cup Y$ with $|X| = j$ such that if $x \in X$, $y \in Y$, then $x < y$ and, if K is any $(k + 1)$-subset of H with $K \cap X \neq \emptyset$, then the color of K is determined only by $\min_{x \in K} x$.

We prove this by induction on j. For $j = 0$ it is trivial and we can choose $f(0, k) = k$. Assume that it holds for some $j \geq 0$. Let $f(j + 1, k) = f(j, n(k, k - j - 1, r) + j + 1)$, let $n \geq f(j + 1, k)$ and suppose the $(k + 1)$-subsets of $[1, n]$ are arbitrarily r-colored. By the induction hypothesis, there is a subset $T = X' \cup Y'$ with $|X'| = j$ which satisfies the required conditions. Let $x_{j+1} = \min_{y \in Y'} y$.

We now r-color the k-subsets of $Y' - \{x_{j+1}\}$ by assigning to the k-set $Z \subseteq Y' - \{x_{j+1}\}$ the same color that the $(k + 1)$-set $Z \cup \{x_{j+1}\}$ has. By the definition of $n(k, k - j - 1, r)$, there is a $(k - j - 1)$-subset Y of $Y' - \{x_{j+1}\}$ with all its k-subsets having the same color. Hence, if $X = X' \cup \{x_{j+1}\}$, then $X \cup Y$ satisfies the required conditions for the value $j + 1$ and the induction step is complete.

Now, consider an $n \geq f((l - 1)r + 1, (l - 1)r + 1)$ and let the $(k + 1)$-subsets of $[1, n]$ be r-colored. By the definition of f, there is an $((l - 1)r + 1)$-subset X of $[1, n]$ such that the color of any $(k + 1)$-subset of X is determined only by its least element. Hence, each element determines a color and since $|X| = (l - 1)r + 1$, there is an l-subset L of X with all $(k + 1)$-subsets of L having the same color. This completes the induction step and the theorem is proved. ∎

THEOREM (van der Waerden): *Given k and r, there exists an integer $W(k, r)$ such that any r-coloring of $[1, W(k, r)]$ must contain an arithmetic progression of k terms all having the same color.*

Proof [17]: For a positive integer l, let us call two m-tuples $(x_1, \ldots, x_m)$, $(x_1', \ldots, x_m') \in [0, l]^m$ *l-equivalent* if they agree up through their last occurrences of l. For any $l, m \geq 1$, consider the statement

$S(l, m)$: For any r, there exists $W(l, m, r)$ so that for any function $C:[1, W(l, m, r)] \to [1, r]$, there exist positive $a, d_1, \ldots, d_m$ such that $C(a + \sum_{i=1}^{m} x_i d_i)$ is constant on each l-equivalence class of $[0, l]^m$.

Fact 1. $S(l, m)$ for some $m \geq 1 \Rightarrow S(l, m + 1)$.

Proof: For a fixed r, let $M = W(l, m, r)$, $M' = W(l, 1, r^M)$ and suppose $C:[MM'] \to [1, r]$ is given. Define $C':[1, M'] \to [1, r^M]$ so that $C'(k) = C'(k')$ iff $C(kM - j) = C(k'M - j)$ for all $0 \leq j < M$. By the inductive hypothesis, there exist a' and d' such that $C'(a' + xd')$ is constant for $x \in [0, l - 1]$. Since $S(l, m)$ can apply to the interval $[a'M + 1, (a' + 1)M]$, then by the choice of M, there exist $a, d_1, \ldots, d_m$ with all sums $a + \sum_{i=1}^{m} x_i d_i$, $x_i \in [0, l]$, in $[a'M + 1, (a' + 1)M]$ and with $C(a + \sum_{i=1}^{m} x_i d_i)$ constant on l-equivalence classes. Set $d_i' = d_i$ for $i \in [1, m]$ and $d_{m+1}' = d'M$; then $S(l, m + 1)$ holds.

Fact 2. $S(l, m)$ for all $m \geq 1 \Rightarrow S(l + 1, 1)$.

Proof: For a fixed r, let $C:[1, 2N(l, r, r)] \to [1, r]$ be given. Then there exist $a, d_1, \ldots, d_r$ such that for $x_i \in [0, l]$, $a + \sum_{i=1}^{r} x_i d_i \le W(l, r, r)$ and $C(a + \sum_{i=1}^{r} x_i d_i)$ is constant on l-equivalence classes. By the box principle there exist $u < v$ in $[0, r]$ such that $C(a + \sum_{i=1}^{u} l d_i) = C(a + \sum_{i=1}^{v} l d_i)$. Therefore $C((a + \sum_{i=1}^{u} l d_i) + x(\sum_{i=u+1}^{v} d_i))$ is constant for $x \in [0, l]$. This proves $S(l + 1, 1)$.

Since $S(1, 1)$ holds trivially, then by induction $S(l, m)$ is valid for all $l, m \ge 1$. Van der Waerden's theorem is $S(l, 1)$. ∎

THEOREM (Schur): *For all r there is an integer $N(r)$ such that any r-coloring of $[1, N(r)]$ contains three elements x, y, z having the same color and which satisfy $x + y = z$.*

Proof: Choose $N(r) = n(2, 3, r)$, the Ramsey number from Ramsey's Theorem. Any r-coloring of $[1, N(r)]$ induces an r-coloring of the edges of $K_{N(r)}$ by assigning to the edge $\{i, j\}$ the color that $|i - j|$ has. By the definition of $N(r)$, there exists a monochromatic triangle in $K_{N(r)}$, i.e., $x < y < z$ such that $z - x$, $z - y$, $y - x$ all have the same color. But $(z - y) + (y - x) = z - x$ so we are done. ∎

THEOREM (Greenwood and Gleason): $r(K_3, K_3, K_3) = 17$.

Proof: Form a 3-coloring of K_{16} by labelling the vertices with the elements of $GF(16)$ and coloring the edge $\{x, y\}$, $x, y \in GF(16)$, according to the coset of the group of cubic residues in which the difference $x - y$ lies. This coloring is well defined since $-1 \equiv 1$ (mod 2). It is not hard to check that no monochromatic triangle is formed and so $r(K_3, K_3, K_3) > 16$.

To show that $r(K_3, K_3, K_3) \le 17$, let K_{17} be arbitrarily 3-colored. For a fixed vertex x, some color, say blue, occurs in at least six edges incident to x. If the vertices at the other ends of these six blue edges span a blue edge then we have a blue triangle. On the other hand, if no blue edge is so spanned, then we have a 2-colored K_6. Since $r(K_3, K_3) = 6$, then we have a monochromatic K_3 in this case as well. Thus $r(K_3, K_3, K_3) \leq 17$ and the proof is completed. ∎

REFERENCES

0. J. E. Baumgartner, "A short proof of Hindman's theorem", *J. Combinatorial Theory* (A), **17** (1974), 384-386.
1. E. Berlekamp, "A construction for partitions which avoid long arithmetic progressions", *Canad. Math. Bull.*, **11** (1968), 409-414.
2. S. Burr, "Generalized Ramsey numbers for graphs—a survey", *Graphs and Combinatorics*, R. Bari and F. Harary, eds., Springer-Verlag, Berlin, 1974.
3. M. Cates and N. Hindman, "Partition theorems for subspaces of vector spaces", *J. Combinatorial Theory* (A), **18** (1975), 1-13.
4. V. Chvatál, "Some unknown van der Waerden numbers", *Combinatorial Structures and Their Applications*, Gordon and Breach, New York, 1970, 31-33.
5. W. Deuber, "Partitionen und Lineare Gleichungsysteme", *Math. Z.*, **133** (1973), 109-123.
6. ____, "Generalizations of Ramsey's theorem", *Proc. Conf. on Infinite and Finite Sets*, Keszthely, 1973, János Bolyai Mathematical Society, Budapest and North-Holland, Amsterdam-London, 1975, vol. 1, 323-332.
7. B. Dushnik and E. W. Miller, "Partially ordered sets", *Amer. J. Math.*, **63** (1941), 600-610.
8. P. Erdös and G. Szekeres, "On a combinatorial problem in geometry", *Compositio Math.*, **2** (1935), 463-470.
9. P. Erdös, "Some remarks on the theory of graphs", *Bull. Amer. Math. Soc.*, **53** (1947), 292-294.
10. P. Erdös and A. Hajnal, "Unsolved problems in set theory", *Proc. Sympos. in Pure Math.*, 13 pt. 1, Amer. Math. Soc., Providence, 1971, 17-48.
11. P. Erdös, R. L. Graham, P. Montgomery, B. L. Rothschild, J. Spencer, and E. G. Straus, "Euclidean Ramsey theorems I", *J. Combinatorial Theory* (A), **14** (1973), 341-363.
12. ____, "Euclidean Ramsey Theorems II, III", *Proc. Conf. on Infinite and Finite Sets*, Keszthely, North-Holland, Amsterdam, 1973, 323-332.
13. J. H. Folkman, "Graphs with monochromatic complete subgraphs in every edge coloring", *SIAM J. Appl. Math.*, **18** (1970), 19-24.
14. R. L. Graham, K. Leeb, B. L. Rothschild, "Ramsey's theorem for a class of categories", *Advances in Math.*, **8** (1972), 417-433 (Errata, *ibid.* **10** (1973), 326-7).
15. R. L. Graham and B. L. Rothschild, "Ramsey's theorem for n-parameter sets", *Trans. Amer. Math. Soc.*, **159** (1971), 257-292.
16. ____, "A survey of finite Ramsey theorems", *Proc. of 2nd Louisiana Conf. on Combinatorics, Graph Theory and Computing*, 1971.

17. ____, "A short proof of van der Waerden's theorem on arithmetic progressions", *Proc. Amer. Math. Soc.*, **42** (1974), 385-6.
18. J. E. Graver and J. Yackel, "Some graph theoretic results associated with Ramsey's theorem", *J. Combinatorial Theory*, **4** (1968), 125-175.
19. R. E. Greenwood and A. M. Gleason, "Combinatorial relations and chromatic graphs", *Canad. J. Math.*, (1955), 1-7.
20. N. Hindman, "Finite sums from sequences within cells of a partition of **N**", *J. Combinatorial Theory* (A), **17** (1974), 1-11.
21. K. Leeb, "The categories of combinatorics", *Combinatorial Structures and Their Applications*, Gordon and Breach, New York, 1970.
22. ____, "Vorlesungen über Pascaltheorie", *Lecture Notes*, Universität Erlangen, 1973.
23. J. Nešetřil and V. Rödl, "The Ramsey property for graphs with forbidden complete subgraphs", *J. Combinatorial Theory* (B), **20** (1976), 243-249.
24. J. Nešetřil and V. Rödl, "Partitions of finite relations and set systems", *J. Combinatorial Theory* (A), **22** (1977), 289-312.
25. R. Rado, "Studien zur Kombinatorik", *Math Zeit.*, **36** (1933), 424-480.
26. ____, "Note on combinatorial analysis", *Proc. London Math. Soc.*, (2) **48** (1943), 122-160.
27. F. P. Ramsey, "On a problem in formal logic", *Proc. London Math. Soc.*, (2) **30** (1930), 264-286.
28. I. Schur, "Uber die Kongruenz $x^m + y^m \equiv z^m \pmod{p}$", *Jber. Deutsch. Math. Verein.*, **25** (1916), 114-116.
29. E. Szemerédi, "On sets of integers containing no k elements in arithmetic progression", *Acta Arith.*, **27** (1975), 199-245.
30. B. L. van der Waerden, "Beweis einer Baudetschen Vermutung", *Nieuw Arch. Wisk.*, **15** (1927), 212-216.
31. J. Yackel, "Inequalities and asymptotic bounds for Ramsey numbers", *J. Combinatorial Theory* (B), **12** (1972), 56-68.

GENERATING FUNCTIONS*

Richard P. Stanley

I. ENUMERATION

One of the fundamental concepts in combinatorial theory is that of *enumeration*, and one of the basic techniques for dealing with problems of enumeration is that of *generating functions*. In this paper we shall survey some of the highlights of the theory of generating functions and shall discuss some applications to specific problems of enumeration. Many examples will be given—some classical and well-known, some more obscure, and a few new. Our object will be two-fold: (1) impart to the casual reader some of the flavor of recent work with generating functions, and (2) impart some facility for using generating functions as a tool for solving combinatorial problems. In some instances, such as Proposition 4.13, Proposition 5.3 and Example 6.11, we bring the reader near the frontiers of what we believe to be exciting new areas of research.

---

*Partially supported by NSF Grant No. P36739.

Naturally, we can only give a small selection of topics from the vast subject of the enumerative theory of generating functions. In Section 3 we shall consider the "abstract" theory of generating functions. In Sections 4 and 5 we shall be concerned with two special classes of generating functions—rational functions and algebraic functions. Finally in Section 6 we shall discuss a result, known as the "exponential formula", which deals with the occurrence of the exponential function in certain types of enumeration problems. Further information about generating functions can be obtained from, e.g., [10], [28], [33]. These books are devoted almost entirely to the use of generating functions for solving combinatorial problems.

Let I be an index set, and let $\mathcal{S} = \{S_\iota : \iota \in I\}$ be a system of finite sets S_ι indexed by I. For our purposes, the *fundamental problem of enumeration* is to "determine" the cardinality of each S_ι as a function of $\iota \in I$. Equivalently, we wish to determine the *counting function* $N: I \to \mathbf{N}$ defined by $N(\iota) = |S_\iota|$. Here $\mathbf{N}$ denotes the set of nonnegative integers and $|S_\iota|$ denotes the cardinality (number of elements) of S_ι. In any combinatorial problem, there will be some combinatorial relationship between $\iota \in I$ and S_ι. For instance, we could have $I = \mathbf{N}$ with S_n being the set of all subsets of the set $[n] = \{1, 2, \ldots, n\}$. Here $N(n) = 2^n$. Examples of index sets frequently encountered in enumeration problems include the following:

(i) $\mathbf{N}$, the nonnegative integers.

(ii) $\mathbf{N} \times \mathbf{N}$, pairs (k, n) of nonnegative integers. For instance, $S_{(k, n)}$ could be all subsets of $[n]$ of cardinality k. Then $N(k, n)$ is commonly denoted $\binom{n}{k}$.

(iii) $\mathbf{P}$, the positive integers. For instance, S_n could be the set of divisors of n, so $N(n)$ is the well-known number-theoretic function $d(n)$.

(iv) π, the set of all partitions $\lambda = (\lambda_1, \lambda_2, \ldots)$ of all nonnegative integers. Here $\lambda_i \in \mathbf{N}$, $\lambda_1 \geq \lambda_2 \geq \cdots$, and $\Sigma \lambda_i$ is finite. If $\Sigma \lambda_i = n$, then λ is called a *partition of n*, denoted $\lambda \vdash n$. For instance, S_λ could be the set of all permutations in the symmetric group $\mathfrak{S}_n$ ($\lambda \vdash n$) whose cycles have lengths $\lambda_1, \lambda_2, \ldots$. If we write $\lambda = \langle 1^{r_1} 2^{r_2} 3^{r_3} \cdots \rangle$ to signify that exactly r_i of the λ_j's are equal to

i (so if $\lambda \vdash n$ then $\Sigma\, ir_i = n$), then $N(\lambda) = |S_\lambda| = n!/(1^{r_1}r_1!)(2^{r_2}r_2!)(3^{r_3}r_3)\cdots$.

II. GENERATING FUNCTIONS

We shall not attempt a rigorous general definition of generating functions but shall content ourselves with various examples. Heuristically, a generating function is a representation of a counting function $N\colon I \to \mathbf{N}$ as an element $F(N)$ of some algebra $\mathcal{C}$. The following are examples of types of generating functions which have actually arisen in specific enumeration problems.

2.1. Ordinary generating functions. Here $I = \mathbf{N}$, $\mathcal{C} = \mathbf{C}[[X]]$ (the ring of formal power series over the complex numbers $\mathbf{C}$), and $N\colon \mathbf{N} \to \mathbf{N}$ is represented by

$$F(N; X) = \sum_{n=0}^{\infty} N(n)\, X^n,$$

called the *ordinary generating function* of N. Sometimes $I = \mathbf{P}$ and the sum starts at $n = 1$.

2.2. Exponential generating functions. $I = \mathbf{N}$ and $\mathcal{C} = \mathbf{C}[[X]]$ as before, while

$$F(N; X) = \sum_{n=0}^{\infty} N(n) X^n/n!$$

2.3. Eulerian generating functions (cf. [16]). Let q be a fixed positive integer (almost always taken in practice to be a prime power corresponding to the field $GF(q)$). Take $I = \mathbf{N}$, $\mathcal{C} = \mathbf{C}[[X]]$, and

$$F(N; X) = \sum_{n=0}^{\infty} N(n) X^n/(1+q)(1+q+q^2)\cdots(1+q+\cdots+q^{n-1}).$$

Frequently the denominator is replaced with $(1 - q)(1 - q^2) \cdots (1 - q^n)$; this amounts to the transformation $X \to X/(1 - q)$. One advantage of our "normalization" is that $F(N; X)$ reduces to an exponential generating function when $q = 1$.

2.4. Doubly-exponential generating functions. $I = \mathbf{N}$, $\mathcal{Q} = \mathbf{C}[[X]]$, and

$$F(N; X) = \sum_{n=0}^{\infty} N(n) X^n / (n!)^2.$$

For instance [3], if $N(n)$ is the number of $n \times n$ matrices of nonnegative integers such that every row and column sum equals two, then $F(N; X) = e^{X/2} (1 - X)^{-1/2}$. (See Example 6.11.)

2.5. Chromatic generating functions (cf. [7], [32], [39]). $I = \mathbf{N}$, $\mathcal{Q} = \mathbf{C}[[X]]$, $q \in \mathbf{P}$ is fixed, and

$$F(N; X) = \sum_{n=0}^{\infty} N(n) X^n / q^{\binom{n}{2}} n!.$$

Sometimes one sees $q^{\binom{n}{2}}$ replaced with $q^{n^2/2}$, amounting to the transformation $X \to X/q^{1/2}$.

2.6. Power series in two variables. Here $I = \mathbf{N} \times \mathbf{N}$ (or possibly $\mathbf{P} \times \mathbf{N}$, $\mathbf{N} \times \mathbf{P}$, $\mathbf{P} \times \mathbf{P}$) and $\mathcal{Q} = \mathbf{C}[[X, Y]]$, the ring of formal power series in two variables X, Y over $\mathbf{C}$. Then $F(N; X, Y)$ can take such forms as

$$F(N; X, Y) = \sum_{m=0}^{\infty} \sum_{n=0}^{\infty} N(m, n) X^m Y^n,$$

$$F(N; X, Y) = \sum_{m=0}^{\infty} \sum_{n=0}^{\infty} N(m, n) X^m Y^n / n!,$$

$$F(N; X, Y) = \sum_{m=0}^{\infty} \sum_{n=0}^{\infty} N(m, n) X^m Y^n / m! n!,$$

etc.

2.7. Power series in infinitely many variables. There are two common possibilities for I. One is the set $\mathbf{S}^*$ of all sequences $(n_1, n_2, \ldots)$ of nonnegative integers with only finitely many $n_i \neq 0$, while the other is the set π of all partitions of nonnegative integers. In either case $\mathcal{C} = \mathbf{C}[[X_1, X_2, \ldots]] = \mathbf{C}[[\mathbf{X}]]$, the ring of formal power series in $X_1, X_2, \ldots$ over $\mathbf{C}$ (each monomial containing only finitely many different X_i). If $I = \mathbf{S}^*$, then

$$F(N; X_1, X_2, \ldots) = \sum_{n_1, n_2, \ldots = 0}^{\infty} N(n_1, n_2, \ldots) X_1^{n_1} X_2^{n_2} \ldots,$$

while if $I = \pi$, then

$$F(N; X_1, X_2, \ldots) = \sum_{n=0}^{\infty} \sum_{\lambda \vdash n} N(\lambda) X_1^{r_1} X_2^{r_2} \ldots, \qquad (1)$$

where λ has exactly r_i parts equal to i.

2.8. Dirichlet series. $I = \mathbf{P}$ and $\mathcal{C}$ is the algebra $\mathfrak{D}$ of all formal Dirichlet series with coefficients in $\mathbf{C}$. Then

$$F(N; s) = \sum_{n=1}^{\infty} N(n) n^{-s}.$$

As $\mathbf{C}$-algebras the two algebras $\mathfrak{D}$ and $\mathbf{C}[[\mathbf{X}]]$ are isomorphic (*via* the transformation $X_i \to p_i^{-s}$ where p_i is the ith prime), but of course their analytic behavior is entirely different.

III. BINOMIAL TYPE

The problem arises of trying to "explain" combinatorially why certain types of generating functions such as $\Sigma N(n) X^n$ and $\Sigma N(n) \times$

$X^n/n!$ often arise, while other types like $\Sigma N(n)X^n/(1 + n^2)$ or $\Sigma N(n)X^n/1^1\, 2^2\, 3^3 \cdots n^n$ never seem to occur. Two abstract theories of generating functions have been formulated to try to solve this problem—the Doubilet-Rota-Stanley theory of "reduced incidence algebras" [11], and the Bender-Goldman theory of "prefabs" [7] (cf. also the "dissect" theory of M. Henle [22], which combines features of both the preceding theories). To give the reader some feeling for this subject we shall discuss the main theorem of Doubilet-Rota-Stanley concerning power series generating functions in one variable.

A partially ordered set (or *poset*, for short) P will be said to be *binomial* if it satisfies the following three conditions:

(a) P is locally finite, i.e., every interval $[x, y] = \{z: x \leq z \leq y\}$ is finite, and P contains arbitrarily large finite chains. (A *chain* is a totally ordered subset of P.)

(b) For every interval $[x, y]$ of P, all maximal chains between x and y have the same length $n = n(x, y)$. We then call $[x, y]$ an *n-interval*. (The *length* of a chain is one less than its number of elements.)

(c) For all $n \in \mathbf{N}$, any two n-intervals contain the same number $B(n)$ of maximal chains.

Clearly from these definitions we have $B(0) = B(1) = 1$, $B(2) = |[x, y]| - 2$, where $[x, y]$ is any 2-interval, and $B(0) \leq B(1) \leq B(2) \leq \cdots$.

Examples of binomial posets

3.1: $P = \mathbf{N}$ with the usual order. Then $B(n) = 1$ for all $n \in \mathbf{N}$.

3.2: P is the lattice of all finite subsets of $\mathbf{N}$, ordered by inclusion. Then $B(n) = n!$.

3.3: P is the lattice of all finite-dimensional subspaces of a vector space of infinite dimension over $GF(q)$, ordered by inclusion. Then $B(n) = (1 + q)(1 + q + q^2) \cdots (1 + q + q^2 + \cdots + q^{n-1})$.

3.4: P is the poset of all subsets of $\mathbf{N} \times \mathbf{N}$ of the form $S \times T$, where S and T are finite subsets of $\mathbf{N}$ of the same cardinality, ordered by inclusion. Then $B(n) = n!^2$.

3.5: Let V be an infinite vertex set, let $q \in \mathbf{P}$ be fixed, and let P be the set of all pairs (G, σ), where G is a function from all 2-sets

$\{u, v\} \subset V$ ($u \neq v$) into $\{0, 1, \ldots, q-1\}$ such that all but finitely many values of G are 0, and where $\sigma: V \to \{0, 1\}$ is a map satisfying the two conditions: (a) if $G(\{u, v\}) > 0$ then $\sigma(u) \neq \sigma(v)$, and (b) $\sum_{v \in V} \sigma(v) < \infty$.

If (G, σ) and (H, τ) are in P, define $(G, \sigma) \leq (H, \tau)$ if:
(i) $\sigma(v) \leq \tau(v)$ for all $v \in V$, and
(ii) if $\sigma(u) = \tau(u)$ and $\sigma(v) = \tau(v)$, then $G(\{u, v\}) = H(\{u, v\})$.

Then P is a binomial poset with $B(n) = n! q^{\binom{n}{2}}$. This rather artificial-looking example arises naturally in [39, §3] in connection with the coloring of graphs.

Observe that the numbers $B(n)$ considered in 3.1-3.5 appear in the power series generating functions of 2.1-2.5. If we can somehow associate a binomial poset with generating functions of the form $\Sigma N(n) X^n / B(n)$, then we will have "explained" the form of the generating functions of 2.1-2.5. We also will have provided some justification of the vague metaprinciple that ordinary generating functions are associated with the nonnegative integers, exponential generating functions with sets, Eulerian generating functions with vector spaces, etc.

To see the connection between binomial posets and generating functions of the form $\Sigma N(n) X^n / B(n)$, it is necessary to consider incidence algebras. If P is any locally finite poset, the *incidence algebra* $I(P)$ of P (over $\mathbf{C}$, say) is the vector space of all functions $f: S(P) \to \mathbf{C}$, where $S(P)$ is the set of all nonvoid intervals $[x, y]$ of P, endowed with the multiplication (convolution)

$$fg(x, y) = \sum_{z \in [x, y]} f(x, z) g(z, y).$$

(We write $f(x, y)$ for $f([x, y])$, etc.) Note that the above sum is finite since P is locally finite. It is easily seen that $I(P)$ is an associative algebra with identity δ given by $\delta(x, y) = \delta_{xy}$ (the Kronecker delta). If P is binomial, let $R(P)$ be the subspace of $I(P)$ consisting of functions f constant on n-intervals, i.e., $f(x, y) = f(z, w)$ whenever $[x, y]$ and $[z, w]$ have the same length. If $f \in R(P)$, we write $f(n)$ for $f(x, y)$, where $[x, y]$ is an n-interval.

A fundamental property of binomial posets is that $R(P)$ is a subalgebra of $I(P)$, i.e., $R(P)$ is closed under convolution. Note also that $\delta \in R(P)$. Indeed, it is easy to see that

$$fg(n) = \sum_{i=0}^{n} \begin{bmatrix} n \\ i \end{bmatrix} f(i)g(n-i), \qquad (2)$$

where $\begin{bmatrix} n \\ i \end{bmatrix}$ denotes the number of elements z in an n-interval $[x, y]$ such that $[x, z]$ is an i-interval. Since $B(i)B(n-i)$ maximal chains of $[x, y]$ pass through a given such z, we have

$$\begin{bmatrix} n \\ i \end{bmatrix} = \frac{B(n)}{B(i)B(n-i)}. \qquad (3)$$

This is the P-analogue of the formula

$$\binom{n}{i} = \frac{n!}{i!(n-i)!}.$$

This analogy is strengthened further by observing that

$$B(n) = A(n)A(n-1) \cdots A(1),$$

where $A(i) = \begin{bmatrix} i \\ 1 \end{bmatrix}$.

We immediately have from (2) and (3) the following main theorem on binomial posets.

3.6. THEOREM: *Let P be a binomial poset. Then $R(P)$ is isomorphic to $\mathbf{C}[[X]]$ via*

$$f \mapsto F_f(X) = \sum_{n=0}^{\infty} f(n)X^n/B(n).$$

Let us consider some applications. In the following example, P is assumed to be a binomial poset:

3.7. Example: Define $\zeta \in R(P)$ by $\zeta(n) = 1$ for all $n \in \mathbf{N}$. Then for an n-interval $[x, y]$,

$$\zeta^2(n) = \zeta^2(x, y) = \sum_{z \in [x,y]} \zeta(x, z)\zeta(z, y)$$

$$= \sum_{z \in [x,y]} 1 = |[x, y]|.$$

Hence, the cardinality $N(n)$ of an n-interval is given by

$$\sum_{n=0}^{\infty} N(n)X^n/B(n) = \left(\sum_{n=0}^{\infty} X^n/B(n)\right)^2.$$

Thus from 3.1 we have that the cardinality $N(n)$ of a chain of length n satisfies

$$\sum_{n=0}^{\infty} N(n)X^n = \left(\sum_{n=0}^{\infty} X^n\right)^2 = 1/(1 - X)^2 = \sum_{n=0}^{\infty} (n + 1)X^n,$$

whence $N(n) = n + 1$. Similarly from 3.2 the number $N(n)$ of subsets of an n-element set satisfies

$$\sum_{n=0}^{\infty} N(n)X^n/n! = \left(\sum_{n=0}^{\infty} X^n/n!\right)^2 = e^{2X} = \sum_{n=0}^{\infty} 2^n X^n/n!,$$

whence $N(n) = 2^n$. The analogous formula for Eulerian generating functions first appeared in [16].

3.8. Example: For $n \geq 1$, let $N(n)$ be the number of sequences $0 = a_0 < a_1 < \cdots < a_k = n$ of integers a_i such that no $a_{i+1} - a_i = 1$ for $0 \leq i < k$. Also set $N(0) = 1$. Let $P = \mathbf{N}$, and define $\eta \in R(P)$ by

$$\eta(n) = \begin{cases} 0, & n = 0 \text{ or } 1, \\ 1, & n \geq 2. \end{cases}$$

The number of sequences we seek of length k is clearly $\eta^k(n)$, so

$$N(n) = \left(\sum_{k=0}^{\infty} \eta^k\right)(n) = (1 - \eta)^{-1}(n).$$

Now $F_\eta(X) = X^2 + X^3 + \cdots = X^2/(1 - X)$, so

$$\sum_{n=0}^{\infty} N(n)X^n = (1 - F_\eta(X))^{-1}$$

$$= (1 - X)/(1 - X - X^2) \qquad (4)$$

$$= 1 + X^2 + X^3 + 2X^4 + 3X^5 + \cdots.$$

It follows that $N(n + 1) = F_n$, the nth Fibonacci number, a well-known result. The reader should by now be able to find analogous results for sets, vector spaces, etc., and invent his own modifications and generalizations. For instance, if $|S| = n$ and $M(n)$ denotes the number of chains $\emptyset = S_0 \subset S_1 \subset \cdots \subset S_k = S$ such that each $|S_{i+1} - S_i| \geq 2$, $0 \leq i < k$, then in complete analogy to (4) we have

$$\sum_{n=0}^{\infty} M(n)X^n/n! = (1 - (e^X - 1 - X))^{-1} = (2 + X - e^X)^{-1}.$$

For a host of other applications and generalizations to other types of generating functions (such as Dirichlet series), see [11].

IV. RATIONAL FUNCTIONS IN ONE VARIABLE

Theorem 3.6 sheds considerable light on the "meaning" of generating functions and reduces certain simple types of combinatorial problems to a routine computation. However, it does not seem worthwhile to attack more complicated problems from this point of view. For the remainder of this paper we will consider other techniques for obtaining and analyzing generating functions. In this section we will consider some aspects of ordinary generating functions

$$F(X) = \sum_0^\infty N(n) X^n$$

which are rational functions in the ring $\mathbf{C}[[X]]$, i.e., for which there exist polynomials $P(X)$, $Q(X) \in \mathbf{C}[X]$ such that $F(X) = P(X)Q(X)^{-1}$. Here $Q(X)^{-1}$ is interpreted to be the element of $\mathbf{C}[[X]]$ satisfying $Q(X)Q(X)^{-1} = 1$. $Q(X)^{-1}$ will exist if and only if $Q(0) \neq 0$. The fundamental property of rational functions in $\mathbf{C}[[X]]$ from the viewpoint of enumeration is the following:

4.1. THEOREM: *Let $\alpha_1, \alpha_2, \ldots, \alpha_d$ be a fixed sequence of complex numbers, $d \geq 1$ and $\alpha_d \neq 0$. The following conditions on a function $N: \mathbf{N} \to \mathbf{C}$ are equivalent:*

(i) $\sum_{n=0}^\infty N(n)X^n = P(X)/Q(X)$,

where $Q(X) = 1 + \alpha_1 X + \alpha_2 X^2 + \cdots + \alpha_d X^d$ and $P(X)$ is a polynomial in X, of degree less than d, relatively prime to $Q(X)$.

(ii) *For all $n \geq 0$,*

$$N(n + d) + \alpha_1 N(n + d - 1) + \alpha_2 N(n + d - 2) \\ + \cdots + \alpha_d N(n) = 0, \qquad (5)$$

and N satisfies no relation $N(n + c) + \beta_1 N(n + c - 1) + \cdots + \beta_c N(n) = 0$, where $c < d$ and each β_i is a fixed element of $\mathbf{C}$.

(iii) *For all $n \geq 0$,*

$$N(n) = \sum_{i=1}^k P_i(n)\gamma_i{}^n,$$

where $1 + \alpha_1 X + \alpha_2 X^2 + \cdots + \alpha_d X^d = \prod_{i=1}^k (1 - \gamma_i X)^{d_i}$, the γ_i's are distinct, and $P_i(n)$ is a polynomial in n of degree $d_i - 1$. □

Theorem 4.1 is well known in the calculus of finite differences and has many proofs. Perhaps the simplest proof involves decomposing $P(X)/Q(X)$ by partial fractions. Other proofs can be given using the calculus of residues, finite difference operators, the method of undetermined coefficients, etc.

The main application of Theorem 4.1 is as follows: One frequently can show by non-combinatorial means that a generating function $\Sigma N(n) X^n$ is rational. Theorem 4.1(ii) and (iii) then provide a simple recurrence for calculating $N(n)$ and a means of estimating the growth of $N(n)$. We shall give as a non-trivial illustration a minor modification of a result of D. Klarner [23] and G. Pólya [46].

4.2. Example: A *polyomino* is a finite union P of unit squares in the plane such that the vertices of the squares have integer coordinates, and P is connected and has no finite cut set. Two polyominoes will be considered *equivalent* if there is a translation which transforms one into the other (reflections and rotations not allowed). Let $N(n)$ be the number of inequivalent n-square polyominoes P with the property that each "row" of P is an unbroken line of squares, i.e., if L is any line segment parallel to the x-axis with its two endpoints in P, then $L \subseteq P$. By convention set $N(0) = 0$. Then $N(1) = 1$, $N(2) = 2$, $N(3) = 6$, etc. It is easily seen that

$$N(n) = \Sigma(n_1 + n_2 - 1)(n_2 + n_3 - 1) \cdots (n_{s-1} + n_s - 1), \quad (6)$$

where the sum is over all ordered partitions $n_1 + n_2 + \cdots + n_s = n$ of n into positive integers n_i (by convention, the partition with $s = 1$ contributes 1 to the sum). Let $N_r(n)$ be the sum of those terms of (6) with $n_1 = r$, where we set $N_n(n) = 1$, and where we set $N_r(n) = 0$ if $r > n$ or $n < 0$. Thus

$$N(n) = \sum_{r=1}^{\infty} N_r(n),$$

$$N_r(n) = \sum_{i=1}^{\infty} (r + i - 1) N_i(n - r), \quad r < n. \quad (7)$$

Define the generating function

$$F(X, Y) = \sum_{n=1}^{\infty} \sum_{r=1}^{\infty} N_r(n) X^r Y^n,$$

so

$$F(1, Y) = \sum_{n=1}^{\infty} N(n) Y^n.$$

Now (7) implies

$$F(X, Y) = \sum_{n=1}^{\infty} X^n Y^n + \sum_{n=1}^{\infty} \sum_{r=1}^{\infty} \sum_{i=1}^{\infty} (r + i - 1) N_i(n - r) X^r Y^n =$$
(8)
$$\frac{XY}{1 - XY} + \frac{X^2 Y^2}{(1 - XY)^2} F(1, Y) + \frac{XY}{1 - XY} \sum_{n=1}^{\infty} \sum_{i=1}^{\infty} i N_i(n) Y^n,$$

by straightforward computation.

Let **D** be the subalgebra of $\mathbf{C}[[X, Y]]$ consisting of all power series $\sum_i \sum_j A_{ij} X^i Y^j$ such that for each $j \in \mathbf{N}$, only finitely many A_{ij} are unequal to 0. Define two linear operators $L_1, L_2: \mathbf{D} \to \mathbf{C}[[Y]]$ as follows:

$$L_1(\sum_i \sum_j A_{ij} X^i Y^j) = \sum_j (\sum_i A_{ij}) Y^j,$$

$$L_2(\sum_i \sum_j A_{ij} X^i Y^j) = \sum_j (\sum_i i A_{ij}) Y^j.$$

Note that L_1 and L_2 have the "representations"

$$L_1 H(X, Y) = H(1, Y), \quad L_2 H(X, Y) = \frac{\partial}{\partial X} H(X, Y)\Big|_{X=1};$$

however, for purposes of generalization it is convenient to regard L_1 and L_2 merely as "abstract" operators.

Define $G(Y) = L_2 F(X, Y)$. By applying L_1 and L_2 to (8), we obtain two linear equations involving $F(1, Y)$ and $G(Y)$. Specifically, we get:

$$F(1, Y) = \frac{Y}{1-Y} + \frac{Y^2}{(1-Y)^2} F(1, Y) + \frac{Y}{1-Y} G(Y)$$

$$G(Y) = \frac{Y}{(1-Y)^2} + \frac{2Y^2}{(1-Y)^3} F(1, Y) + \frac{Y}{(1-Y)^2} G(Y). \qquad (9)$$

Here we have used the easily verified formulas

$$L_2\left(\frac{XY}{1-XY}\right) = Y/(1-Y)^2 \text{ and}$$

$$L_2\left(\frac{X^2 Y^2}{(1-XY)^2}\right) = 2Y^2/(1-Y)^3.$$

Eliminating $G(Y)$ from (9) allows us to solve for $F(1, Y)$ as a function of Y. The final result is

$$F(1, Y) = \frac{Y(1-Y)^3}{1 - 5Y + 7Y^2 - 4Y^3}$$

$$= \frac{1}{16}\left(-5 + 4Y + \frac{5 - 13Y + 7Y^2}{1 - 5Y + 7Y^2 - 4Y^3}\right).$$

Hence we see that

$$N(n+3) = 5N(n+2) - 7N(n+1) + 4N(n), \quad n \geq 2.$$

This recursion is by no means apparent, and no combinatorial proof of it is known.

It is evident that the above method (due essentially to D. Klarner [23], [24], who uses a certain integral representation of our operator L_2) will extend to a much wider class of problems. See also [46]. For instance, the above method yields after a tedious computation the following result:

4.3. Proposition: *Define*

$$N(n) = \Sigma(f_{n_1} + f_{n_2} + f_{n_3})(f_{n_2} + f_{n_3} + f_{n_4}) \cdots (f_{n_{s-2}} + f_{n_{s-1}} + f_{n_s}),$$

where f is any function $f\colon \mathbf{P} \to \mathbf{C}$ and where the sum is over all ordered partitions $n_1 + n_2 + \cdots + n_s = n$ of n ($n_i \geq 1$). By convention, a summand with $s = 1$ is 0 and with $s = 2$ is 1. Define

$$F(X) = \sum_{n=1}^{\infty} N(n)X^n, \quad f = \sum_{n=1}^{\infty} f_n X^n, \quad A = X/(1-X).$$

Let $*$ denote Hadamard product, i.e., $(\Sigma a_n X^n) * (\Sigma b_n X^n) = \Sigma a_n b_n X^n$. Then

$$F(X) = \frac{A^2}{(1-f)^2(1-f-f^2) - 2A(f*f)(1-f^2) - A^2((f*f)^2 + f*f*f)}. \quad \square$$

In obtaining the above expression for $F(X)$ an enormous amount of cancellation takes place. This leads one to suspect that there is some simpler alternative method for obtaining such results. We do not, however, know of such a method.

Theorem 4.1 allows us to deduce the linear recurrence (5) which $N(n)$ satisfies from its generating function $P(X)/Q(X)$. We therefore ask what other properties of $N(n)$ can be "read off" from $P(X)/Q(X)$. A simple and elegant result along these lines has been given by Popoviciu [30] (cf. also [12], [41]). If we are given a function $N\colon \mathbf{N} \to \mathbf{C}$ satisfying a recurrence (5), then clearly there is a unique way of extending N to all of $\mathbf{Z}$ (the integers) such that (5) holds for all $n \in \mathbf{Z}$. Popoviciu's theorem relates the functions $N(n)$ and $N(-n)$. It is easily proved, e.g., by partial fractions.

4.4 Theorem: *Let $N\colon \mathbf{Z} \to \mathbf{C}$ satisfy (5) for all $n \in \mathbf{Z}$. Define*

$$F(X) = \sum_{n=0}^{\infty} N(n)X^n, \quad \overline{F}(X) = \sum_{n=1}^{\infty} N(-n)X^n.$$

Then $F(X)$ and $\overline{F}(X)$ are rational functions of X satisfying $\overline{F}(X) = -F(1/X)$. □

POLYNOMIALS

As important class of functions satisfying a recurrence (5) are the polynomials. In fact, we have the following corollary to Theorem 4.1:

4.5 COROLLARY: *The following conditions on a function N: $\mathbf{N} \to \mathbf{C}$ are equivalent:*

(i) $\sum_{n=0}^{\infty} N(n)X^n = P(X)/(1 - X)^{d+1}$, (10)

where $P(X)$ is a polynomial in X of degree at most d such that $P(1) \neq 0$.

(ii) *For all $n \geq 0$,*

$$\sum_{i=0}^{d+1} (-1)^i \binom{d+1}{i} N(n + i) = 0,$$

while for some $n \geq 0$,

$$\sum_{i=0}^{d} (-1)^i \binom{d}{i} N(n + i) \neq 0.$$

(iii) *$N(n)$ is a polynomial in n of degree d.* □

When a polynomial $N(n)$ arises combinatorially, frequently the coefficients of $P(X)$ (given by (10)) have a combinatorial significance. Moreover, Theorem 4.4 may give useful information about $P(X)$ *via* the following corollary:

4.6. COROLLARY: *Let N: $\mathbf{Z} \to \mathbf{C}$ be a polynomial of degree d,* and let $\sum_{n=0}^{\infty} N(n)X^n = P(X)/(1 - X)^{d+1}$, where $P(X) = a_0 + a_1 X + \cdots + a_d X^d$.

(i) *Define r to be the greatest integer such that $N(0) = N(1) = \cdots = N(r) = 0$. (If $N(0) \neq 0$, let $r = -1$.) If $r \neq -1$, then r is the greatest integer such that $a_0 = a_1 = \cdots = a_r = 0$. Moreover, $N(r + 1) = a_{r+1}$ whatever the value of r.*

(ii) *Define s to be the greatest integer such that $N(-1) = N(-2) = \cdots = N(-s) = 0$. (If $N(-1) \neq 0$, let $s = 0$.) If $s \neq 0$, then s is the greatest integer such that $a_d = a_{d-1} = \cdots = a_{d-s+1} = 0$. Moreover, $N(-s - 1) = (-1)^d a_{d-s}$ whatever the value of s.*

(iii) *Let r and s be given by (i) and (ii). Then $P(X) = X^{d+1+r-s} P(1/X)$ if and only if $N(n) = (-1)^d N(r - s - n)$ for all $n \in \mathbf{Z}$.*

(iv) *The leading coefficient of $N(n)$ is $P(1)/d!$.* □

PARTITIONS AND PERMUTATIONS

The theory of partitions is a highly developed, elegant, and extensive branch of combinatorics. It originated with Euler in 1748 and has occupied the attention of many eminent researchers, such as Jacobi, Sylvester, Hardy and Littlewood, and MacMahon. For an introduction to this subject, see for example [20, Ch. 19], [4, Chs. 12-14], [1], [5], [44]. Generating functions have proved to be an invaluable tool in the study of partitions. We have space here to consider only a very small part of the subject, one in which rational generating functions play an important role. This is the subject of *P-partitions*, various aspects of which were considered by MacMahon, Bender, Knuth, Gordon, Kreweras, E. M. Wright, and others, with a general development first appearing in [37].

Let P be a finite partially ordered set of cardinality p. A *P-partition* of $n \in \mathbf{N}$ is an order-reversing map $\sigma: P \to \mathbf{N}$ satisfying $\sum_{x \in P} \sigma(x) = n$. The statement that σ is *order-reversing* means $\sigma(x) \geq \sigma(y)$ when $x \leq y$ in P. We say that σ is *strict* if $\sigma(x) > \sigma(y)$ when $x < y$ in P. If for instance P is a p-element chain, then a P-partition of n is equivalent to an ordinary partition of n into at most p parts, as defined in Section 1. If on the other extreme P is a disjoint union of p points, then a P-partition of n is equivalent to a *composition* (ordered partition) of n into p parts, allowing 0 as a part.

Define the following combinatorial concepts associated with P:

$a(n) = $ number of P-partitions of $n \in \mathbf{N}$.
$\bar{a}(n) = $ number of strict P-partitions of $n \in \mathbf{N}$.

$$F(X) = \sum_{n=0}^{\infty} a(n)X^n, \quad \overline{F}(X) = \sum_{n=0}^{\infty} \bar{a}(n)X^n.$$

$\Omega(m) = $ number of P-partitions $\sigma: P \to [m]$.
$\overline{\Omega}(m) = $ number of strict P-partitions $\sigma: P \to [m]$.
$e_s = $ number of *surjective* P-partitions $P \to [s]$.
$\bar{e}_s = $ number of *surjective strict* P-partitions $P \to [s]$.

It is easily seen that if $p \geq 1$, then

$$\Omega(m) = \sum_{s=1}^{p} e_s \binom{m}{s}, \quad \overline{\Omega}(m) = \sum_{s=1}^{p} \bar{e}_s \binom{m}{s},$$

so $\Omega(m)$ and $\overline{\Omega}(m)$ are polynomials in m of degree p and leading coefficient $e_p/p!$.

We shall now establish the connection between P-partitions and permutations. Let $\omega: P \to [p]$ be a fixed *order-preserving bijection* (so $x \leq y$ in P implies $\omega(x) \leq \omega(y)$). Define the *JH-set* $\mathcal{L}$ of P to be the set of all permutations $\pi = (a_1, a_2, \ldots, a_p)$ of $(1, 2, \ldots, p)$ such that if $x < y$ in P, then $\omega(x)$ precedes $\omega(y)$ in π. (The reason for this terminology appears in [38].) Hence $\mathcal{L}$ contains a total of e_p permutations. If $\pi = (a_1, a_2, \ldots, a_p)$ is any permutation of $(1, 2, \ldots, p)$, a *descent* is a pair (a_i, a_{i+1}) such that $a_i > a_{i+1}$, while an *ascent* is such a pair with $a_i < a_{i+1}$. Let $\alpha(\pi)$ (respectively $\bar{\alpha}(\pi)$) be the number of descents (respectively, ascents) of the permutation π. Clearly $\alpha(\pi) + \bar{\alpha}(\pi) = p - 1$. The *greater index* $\iota(\pi)$ of π is defined by

$$\iota(\pi) = \Sigma\{j: a_j > a_{j+1}\}.$$

Similarly, the *lesser index* $\bar{\iota}(\pi)$ is defined by

$$\bar{\iota}(\pi) = \Sigma\{j: a_j < a_{j+1}\}.$$

Hence $\iota(\pi) + \bar{\iota}(\pi) = \binom{p}{2}$. (See [28, Section 104].)

We now state without proof some fundamental results concerning P-partitions. Proofs of more general results may be found in [37], especially Corollary 7.2 and Proposition 13.3.

4.7. Proposition: (i) $F(X)$ and $\overline{F}(X)$ are rational functions of X given explicitly by

$$F(X) = (\sum_{\pi \in \mathcal{L}} X^{\iota(\pi)})/(1 - X)(1 - X^2) \cdots (1 - X^p).$$

$$\overline{F}(X) = (\sum_{\pi \in \mathcal{L}} X^{\bar{\iota}(\pi)})/(1 - X)(1 - X^2) \cdots (1 - X^p).$$

(ii) *We have*

$$\sum_{m=0}^{\infty} \Omega(m) X^m = (X \cdot \sum_{\pi \in \mathcal{L}} X^{\alpha(\pi)})/(1 - X)^{p+1},$$

$$\sum_{m=0}^{\infty} \overline{\Omega}(m) X^m = (X \cdot \sum_{\pi \in \mathcal{L}} X^{\bar{\alpha}(\pi)})/(1 - X)^{p+1}. \quad \square$$

Using the formulas $\alpha(\pi) + \bar{\alpha}(\pi) = p - 1$, $\iota(\pi) + \bar{\iota}(\pi) = \binom{p}{2}$, Theorem 4.4, and the definition of $\mathcal{L}$, we can obtain many interesting corollaries to Proposition 4.7, a sample of which are contained in the following:

4.8. Corollary:
(i) $X^p \overline{F}(X) = (-1)^p F(1/X)$.
(ii) $\overline{\Omega}(m) = (-1)^p \Omega(-m)$.
(iii) *Let* $F(X) = W(X)/(1 - X)(1 - X^2) \cdots (1 - X^p)$. *Then* $W(X)$ *is a monic polynomial with nonnegative integer coefficients of degree* $\binom{p}{2} - \sum_{x \in P} \delta(x)$, *where* $\delta(x)$ *is the length of the longest chain of P with bottom x. Moreover,* $W(0) = 1$ *and* $W(1) = e_p$.

(iv) *Let* $d = \deg W(X)$. *Then* $W(X) = X^d W(1/X)$ *if and only if for each* $x \in P$, *every maximal chain of the sub-partially ordered set* $\{y : y \geq x\}$ *has the same length* $l = l(x)$.

4.9. Example: Let (P, ω) be given by

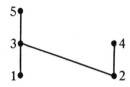

Then $\mathcal{L}$ is given by:

π					$\alpha(\pi)$	$\bar{\alpha}(\pi)$	$\iota(\pi)$	$\bar{\iota}(\pi)$
1	2	3	4	5	0	4	0	10
2	1	3	4	5	1	3	1	9
1	2	4	3	5	1	3	3	7
1	2	3	5	4	1	3	4	6
2	4	1	3	5	1	3	2	8
2	1	4	3	5	2	2	4	6
2	1	3	5	4	2	2	5	5

Hence

$$F(X) = (1 + X + X^2 + X^3 + 2X^4 + X^5)/$$
$$(1 - X)(1 - X^2) \cdots (1 - X^5),$$

$$\overline{F}(X) = (X^5 + 2X^6 + X^7 + X^8 + X^9 + X^{10})/$$
$$(1 - X)(1 - X^2) \cdots (1 - X^5),$$

$$\sum_{m=0}^{\infty} \Omega(m) X^m = (X + 4X^2 + 2X^3)/(1 - X)^6,$$

$$\sum_{m=0}^{\infty} \overline{\Omega}(m) X^m = (2X^3 + 4X^4 + X^5)/(1 - X)^6.$$

4.10. Example: Suppose P is a disjoint union of p points. Then it can be seen directly that

$$\Omega(m) = \overline{\Omega}(m) = m^p,$$

$$F(X) = \overline{F}(X) = 1/(1 - X)^p.$$

Moreover, $\mathcal{L}$ consists of *all* $p!$ permutations of $[p]$, and Proposition 5.3 reduces to classical results on permutations. For instance, the total number of permutations of $[p]$ with s descents is known as an *Eulerian number*, denoted by Knuth [26, Vol. 3, 5.1.3] as $\left\langle \begin{array}{c} p \\ s+1 \end{array} \right\rangle$. Proposition 4.7 implies the well-known result (e.g., [33, pp. 38–39], [26, Vol. 3, 5.1.3, Eq. 8], [10, Ch. 6.5])

$$\sum_{m=0}^{\infty} m^p X^m = \left(\sum_{s=1}^{p} \left\langle \begin{array}{c} p \\ s \end{array} \right\rangle X^s \right) / (1 - X)^{p+1}.$$

Similarly Proposition 4.7 implies that

$$\sum_{\pi} X^{d(\pi)} = (1 + X)(1 + X + X^2) \cdots (1 + X + X^2 + \cdots + X^{p-1}),$$

where the sum is over all permutations π of $[p]$. This remarkable formula is due to MacMahon [27] [26, Vol. 3, 5.1.1].

4.11. Example: Let $P = C_p$, a p-element chain. Then $\mathcal{L}$ consists of the single permutation $(1, 2, \ldots, p)$, and

$$\Omega(m) = \binom{m + p - 1}{p}, \quad \overline{\Omega}(m) = \binom{m}{p},$$

$$F(X) = 1/(1 - X)(1 - X^2) \cdots (1 - X^p),$$

$$\overline{F}(X) = X^{\binom{p}{2}}/(1 - X)(1 - X^2) \cdots (1 - X^p).$$

The formulas for $\Omega(m)$ and $\overline{\Omega}(m)$ are simply the fundamental expressions for counting combinations with or without repetition, while the formulas for $F(X)$ and $\overline{F}(X)$ are basic identities in the theory of partitions.

4.12. Example: Let $P = C_r \times C_s$, a direct (cartesian) product of two chains of cardinalities r and s, where say $r \le s$. It is by no means *a priori* evident that explicit expressions can be given for $\Omega(m)$ and $F(X)$, but such is indeed the case. Namely,

$$\Omega(m) = \frac{\binom{r+m-1}{r}\binom{r+m}{r}\binom{r+m+1}{r}\cdots\binom{r+m+s-2}{r}}{\binom{r}{r}\binom{r+1}{r}\binom{r+2}{r}\cdots\binom{r+s-1}{r}}$$

$$F(X) = 1/(1)(2)^2(3)^3 \cdots (\mathbf{r})^r(\mathbf{r}+1)^r \cdots (\mathbf{s})^r(\mathbf{s}+1)^{r-1}(\mathbf{s}+2)^{r-2} \cdots (\mathbf{r}+\mathbf{s}-1)^1,$$

where $(\mathbf{k}) = 1 - x^k$. These remarkable formulas belong to the fascinating subject of *plane partitions* and are intimately connected with symmetric functions and the representation theory of the symmetric group. For further information, see [36].

The myriad possibilities for modifying or extending the theory of P-partitions remains largely unexplored. As a modest example of what can be done in this direction, we state without proof the following recent result [45].

Let Q_p be the set of all sequences $\pi = (a_1, a_2, \ldots, a_{2p})$ such that each integer $i \in [p]$ appears exactly twice, and such that if $i < j < k$ and $a_i = a_k$, then $a_j > a_i$. It is easily seen that Q_p has cardinality $1 \cdot 3 \cdot 5 \cdots (2p - 1)$. A *descent* of $\pi \in Q_p$ is a pair (a_i, a_{i+1}) with $a_i > a_{i+1}$ $(1 \le i \le 2p - 1)$. Let $s(n, k)$ and $S(n, k)$ denote the Stirling numbers of first and second kinds, respectively. (For a discussion of these numbers, see for example [10, Ch. V].)

4.13. PROPOSITION: *We have the identities*

$$\sum_{n=1}^{\infty} S(n + p, n)X^n = \left(\sum_{i=1}^{p} B_{p,i} X^i\right) / (1 - X)^{2p+1}$$

and

$$(-1)^p \sum_{n=1}^{\infty} s(n + p, n)X^n = \left(\sum_{i=1}^{p} B_{p-i+1, i} X^i\right) / (1 - X)^{2p+1},$$

where $B_{p,i}$ is equal to the number of sequences $\pi \in B_{p,i}$ with exactly $i - 1$ descents. $\square$

For instance, if $p = 2$ then we have (omitting superfluous parentheses and commas) $Q_2 = \{1122, 1221, 2211\}$. Hence

$$\sum_{n=1}^{\infty} S(n + 2, n)X^n = (X + 2X^2)/(1 - X)^5$$

and

$$\sum_{n=1}^{\infty} s(n + 2, n)X^n = (2X + X^2)/(1 - X)^5,$$

agreeing with the known results

$$S(n + 2, n) = \binom{n + 3}{4} + 2\binom{n + 2}{4}$$

and

$$s(n + 2, n) = 2\binom{n + 3}{4} + \binom{n + 2}{4}.$$

V. ALGEBRAIC FUNCTIONS

The elements $F(X)$ of $\mathbf{C}[[X]]$ of the next "level of complexity"

GENERATING FUNCTIONS 123

after the rational functions are the *algebraic functions*. By definition, $Y = F(X)$ is an algebraic function (over $\mathbf{C}$) if there exist polynomials $P_0(X), P_1(X), \ldots, P_d(X) \in \mathbf{C}[X]$ such that

$$P_0(X) + P_1(X)Y + \cdots + P_d(X)Y^d = 0, \qquad (11)$$

as an element of $\mathbf{C}[[X]]$. The least possible d for which (11) holds is the *degree* of Y. If Y satisfies (11), then Y has degree d if and only if $P_d(X) \neq 0$ and $P_0(X) + P_1(X)Y + \cdots + P_d(X)Y^d$ is *irreducible*, considered as a polynomial in Y over the field $\mathbf{C}(X)$.

The theory of algebraic functions has been extensively developed, but most of the results have no direct application to problems of enumeration. We shall discuss some results which *do* apply to enumeration and give several examples to indicate how algebraic functions actually arise in enumeration problems.

5.1. THEOREM (Comtet [9]): *Let* $Y = F(X)$ *be an algebraic function of degree d, given by* $F(X) = \sum\limits_{n=0}^{\infty} N(n)X^n$. *Then there exists a positive integer q and polynomials* $p_0(n), p_1(n), \ldots, p_q(n)$, *such that* $p_q(n) \neq 0$, $\deg p_i(n) < d$, *and for all n sufficiently large,*

$$p_q(n)N(n+q) + p_{q-1}(n)N(n+q-1) + \cdots + p_0(n)N(n) = 0. \qquad (12)$$

Sketch of proof (a streamlined version of Comtet's proof):

Let Y satisfy (11). By differentiating (11) repeatedly with respect to X and using induction, we get that $Y^{(k)} = d^kY/dX^k$ is a rational function $R_k(X, Y)$ of X and Y for all $k \geq 0$. Since Y is algebraic of degree d over $\mathbf{C}(X)$, the functions $1, Y^{(0)} = Y, Y^{(1)}, \ldots, Y^{(d-1)}$ are linearly dependent over $\mathbf{C}(X)$. Write this dependence relation, clear denominators so the coefficient of each $Y^{(k)}$ is a polynomial in X, expand each $Y^{(k)}$ as a power series in X, and equate coefficients of X^n on both sides of the dependence relation to get the desired result. □

5.2. Example: Suppose $2Y^2 - (1 + X)Y + X = 0$, where $Y = \sum_{n=0}^{\infty} N(n)X^n$. Then $Y' = (Y - 1)/(4Y - 1 - X)$ and

$$Y'(X^2 - 6X + 1) - (X - 3)Y + (X - 1) = 0,$$

from which we get

$$(n + 2)N(n + 2) - 3(2n + 1)N(n + 1) + (n - 1)N(n) = 0, \ n \geq 1.$$

5.3. Theorem: *Let* $F(Y, Z) = \sum_{m=0}^{\infty} \sum_{n=0}^{\infty} N(m, n) Y^m Z^n \in \mathbb{C}[[Y, Z]]$. *The diagonal* $D_F(X)$ *of* $F(Y, Z)$ *is the series* $\sum_{n=0}^{\infty} N(n, n) X^n \in \mathbb{C}[[X]]$. *If* $F(Y, Z)$ *is a rational function of* Y *and* Z, *then* $D_F(X)$ *is an algebraic function of* X.

Sketch of proof: Regard $F(Y, Z)$ and $D_F(X)$ as functions of the complex variables X, Y, Z. It is easily seen that $F(Y, Z)$ and $D_F(X)$ converge for X, Y, Z sufficiently small in absolute value (when F is rational). Let C be any sufficiently small circle about the origin in the complex s-plane. We then have for all X sufficiently small (depending on C) in absolute value,

$$D_F(X) = \frac{1}{2\pi i} \int_C F(s, X/s) \frac{ds}{s}.$$

(Cf. [25, Theorem 1] for a rigorous justification of this "formal identity".) By the residue theorem, $D_F(X)$ is equal to the sum of the residues at those poles $s = s(X)$ satisfying $s \to 0$ as $X \to 0$. To compute these residues, write $F(s, X/s) = P(s)/Q(s)$, where $P(s)$ and $Q(s)$ are polynomials in s with coefficients in $\mathbb{C}[X]$. Then the roots of $Q(s)$ are algebraic functions of X. Thus the poles of $F(s, X/s)$ are algebraic functions of X, so the residues at these poles will be rational combinations of these algebraic functions and hence themselves algebraic. From this the proof follows. □

5.4. Example: Let S be a subset of $\mathbf{N} \times \mathbf{N}$ such that $(0, 0) \notin S$. Let $N_S(m, n)$ be the number of ways the vector (m, n) can be written as a sum of vectors belonging to S. The order of summands is taken into account, so for example if

$$S = \{(1, 0), (1, 1), (0, 1)\},$$

then $N_S(1, 1) = 3$, corresponding to $(1, 0) + (0, 1)$, $(0, 1) + (1, 0)$, and $(1, 1)$. Define

$$F_S(Y, Z) = \sum_{m=0}^{\infty} \sum_{n=0}^{\infty} N_S(m, n) Y^m Z^n.$$

It is easily seen that

$$F_S(Y, Z) = 1 \Big/ \Big(1 - \sum_{(i,j) \in S} Y^i Z^j\Big).$$

Hence if S is finite, $F_S(Y, Z)$ is a rational function of Y and Z, so from Theorem 5.3 we obtain:

5.5. Theorem: *Let S be a finite subset of $\mathbf{N} \times \mathbf{N}$ with $(0, 0) \notin S$. Define*

$$G_S(X) = \sum_{n=0}^{\infty} N_S(n, n) X^n.$$

Then $G_S(X)$ is an algebraic function of X, and hence (by Theorem 5.1) *$N(n, n)$ satisfies a recursion of the form* (12) *for n sufficiently large.* □

For an explicit example, take $S = \{(0, 1), (1, 0), (1, 1)\}$. Then $F_S(Y, Z) = 1/(1 - Y - Z - YZ)$ and

$$G_S(X) = \frac{1}{2\pi i} \oint \frac{ds}{s\Big(1 - s - \dfrac{X}{s} - X\Big)}$$

$$= -\frac{1}{2\pi i} \oint \frac{ds}{s^2 + (X - 1)s + X}.$$

The only pole s of the integrand satisfying $s \to 0$ as $X \to 0$ occurs for

$$s = \frac{1 - X - \sqrt{1 - 6X + X^2}}{2}$$

The residue at this pole is

$$G_S(X) = (1 - 6X + X^2)^{-1/2}.$$

For further information on this special case, see [10, Ch. 1, Ex. 21].

It is not necessary for S to be finite in order for $F_S(Y, Z)$ to be rational. For instance, if $S = \mathbf{N} \times \mathbf{N} - \{(0, 0)\}$, then

$$F_S(Y, Z) = 1 \bigg/ \bigg(1 - \bigg(\frac{1}{(1 - Y)(1 - Z)} - 1 \bigg) \bigg),$$

and we obtain

$$G_S(X) = \sum_{n=0}^{\infty} N_S(n, n) X^n = \tfrac{1}{2}[1 + (1 - 12X + 4X^2)^{-1/2}].$$

There is one remaining theorem concerning algebraic functions which is useful in enumeration problems. This is the *Lagrange inversion formula*. Lagrange's formula allows in certain cases an explicit determination of the coefficients of a power series defined by a functional equation, such as an algebraic function. Lagrange's formula is normally stated for analytic functions (e.g., [43, p. 132]), but we shall state a special case valid for formal power series (cf., e.g., [31, Section 5]). We shall use notation from the calculus in a formal way. For instance, if $F(X) = \sum_{n=0}^{\infty} N(n) X^n$, then

$$F(0) = N(0), \quad F'(0) = N(1),$$

$$d^2F/dX^2 = \sum_{n=0}^{\infty} (n + 1)(n + 2) N(n + 2) X^n, \text{ etc.}$$

5.6. Theorem: *Let $\Phi(X) \in \mathbf{C}[[X]]$. There is a unique $Y = Y(X) \in \mathbf{C}[[X]]$ such that*

$$Y = X \cdot \Phi(Y). \tag{13}$$

(Note that the computation of $\Phi(Y)$ does not involve questions of convergence, since from (13) $Y(0) = 0$ and therefore the coefficient of X^n in the expansion of $\Phi(Y)$ is given by a finite sum.) *Let $F(X) \in \mathbf{C}[[X]]$. Then*

$$F(Y) = F(0) + \sum_{n=1}^{\infty} \frac{a_n X^n}{n!},$$

where

$$a_n = \frac{d^{n-1}}{dX^{n-1}} [F'(X)\phi(X)^n]_{X=0}. \square$$

5.7. Example: Let $Y = X + Y^2 + Y^3$, with $Y(0) = 0$. Thus $Y = X/(1 - Y - Y^2)$. If we let $\Phi(X) = 1/(1 - X - X^2)$ and $F(X) = X$, then we have

$$Y = \sum_{n=1}^{\infty} \frac{X^n}{n!} \left[\frac{d^{n-1}}{dX^{n-1}} (1 - X - X^2)^{-n} \right]_{X=0}.$$

Now, $(1 - X - X^2)^{-n} = \sum_{n=0}^{\infty} \binom{-n}{k} (-1)^k (X + X^2)^k$. Hence if $Y = \sum_{n=1}^{\infty} N(n) X^n$, then we easily obtain

$$N(n) = \sum_{\substack{a+2b=n-1 \\ a,\, b \in \mathbf{N}}} \frac{(n + a + b - 1)!}{n!a!b!}. \tag{14}$$

We conclude this section with an example which is a prototype for a wide class of results dealing with planar maps, parenthesization, trees, formal languages, and related topics.

5.8. Example: Let $S \subseteq \mathbf{P} - \{1, 2\}$. Let $N_S(n)$ be the number of ways of dividing a convex $(n + 1)$-gon C into regions R by drawing diagonals not intersecting in the interior of C, such that the number of sides of each region R belongs to S. By convention, $N(0) = 0$, $N(1) = 1$. Fix an edge E of C. Given a decomposition of C enumerated by $N_S(n)$, let k be the number of edges of the region containing E. If we remove E from C, we obtain $k - 1$ new decomposed polygons $C_1, \ldots, C_{k-1}$. If $e(K)$ denotes one less than the number of edges of a polygon K, then $e(C) = e(C_1) + \cdots + e(C_{k-1})$. Hence

$$N_S(n) = \sum_{k \in S} \sum_{\substack{b_1 + \cdots + b_{k-1} = n \\ b_i \in \mathbf{N}}} N_S(b_1) \cdots N_S(b_{k-1}), \quad n \geq 2. \quad (15)$$

Let $Y = F_S(X) = \sum_{n=1}^{\infty} N_S(n) X^n$. Then (15) yields

$$Y = X + \sum_{k \in S} Y^{k-1}. \quad (16)$$

Under certain circumstances Y will be algebraic, e.g., when S is finite, and previous results of this section can be applied.

Suppose, for instance, $S = \{k\}$, $k \geq 3$. Thus, $Y = X + Y^{k-1}$. Theorem 5.6 can be used to show

$$N_S(n) = \begin{cases} 0, \text{ if } (k - 2) \nmid (n - 1), \\ \dfrac{1}{n + t} \binom{n + t}{t}, \text{ if } n - 1 = (k - 2)t. \end{cases}$$

The simple form of this answer suggests that a combinatorial proof might be possible. The expression

$$\frac{1}{n + t} \binom{n + t}{t}$$

is equal to the number of circular permutations of n red beads and t white beads, since $(n, t) = 1$. Thus, we seek an explicit one-to-one correspondence between these circular permutations and the appropriate divisions of a polygon. Such one-to-one correspondences have been described by Raney [31], Tamari [42], and others. These authors prove more generally that the number $L(n)$ of ways of dividing an $(n + 1)$-gon with d diagonals to form a_i regions with $i + 1$ sides (so $d + 1 = \Sigma a_i$ and $n - 1 = \Sigma(i - 1)a_i$) is equal to

$$L(n) = (n + d)!/n! \, \Pi(a_i!). \tag{17}$$

This result was first proved by Etherington and Erdelyi [13] using generating functions. If for example we take $a_2 = a$, $a_3 = b$, and all other $a_i = 0$, we can deduce (14) purely combinatorially. More generally, Raney [31] has shown that (17) is sufficiently general to yield a purely combinatorial proof of Theorem 5.6.

The function Y of (16) can be algebraic without S being finite. For instance, take $S = \{3, 4, 5, \ldots\}$, so $N_S(n)$ is equal to the *total* number of ways of dividing an $(n + 1)$-gon by diagonals not intersecting in the interior. This is the "second problem of Schröder" [35]. By (16), we have

$$Y = X + \sum_{k=3}^{\infty} Y^{k-1} = X + \frac{Y^2}{1 - Y},$$

so $2Y^2 - (1 + {}^{\cdot}X)Y + X = 0$. This gives

$$Y = \frac{1}{4}(1 + X - (1 - 6X + X^2)^{1/2}).$$

This power series Y was the one considered in Example 5.2, so we get

$$(n + 2)N_S(n + 2) - 3(2n + 1)N_S(n + 1) + (n - 1)N_S(n) = 0,$$
$$n \geq 1,$$

as first observed by Comtet [10, p. 57]. (Comtet's formula has a misprint.)

For excellent bibliographies of the many variations of Example 5.8, see [2], [8], or [17]. For the problem of asymptotically estimating the coefficients of an algebraic function, and of asymptotic estimates in general as applied to enumeration, see for example [6]. Finally, we mention that a simpler approach than ours for handling certain types of algebraic functions appears in [26, Vol. 1, Section 2.2.1, Exercises 4 and 11], especially pages 532-534.

VI. THE EXPONENTIAL FORMULA

We wish to explain the ubiquitous appearance in combinatorial enumeration problems of the exponential function. In Section 3 we saw that the exponential function is associated with the incidence algebra of the lattice of finite subsets of $\mathbf{N}$; however, there are many occurrences of the exponential function in combinatorics which cannot be explained in this manner. We will present a general result (Corollary 6.2), which we call the "exponential formula for r-partitions", which leads to a plethora of generating functions involving the exponential function. Although an even more general exponential formula can be given, for simplicity's sake we will restrict ourselves to r-partitions. There are many different approaches to deriving the exponential formula; we choose one which seems to involve the least preparation. A wide variety of examples and special cases will be discussed.

Let S be a finite set with n elements. Recall that a *partition* of S is a collection $\pi = \{B_1, B_2, \ldots, B_k\}$ of non-empty pairwise disjoint subsets B_i of S whose union is S. We say that π is of *type* $(a_1, a_2, \ldots, a_n)$ if exactly a_i of the B_j's have i elements. Thus $\Sigma i a_i = n$ and $\Sigma a_i = k$. We call the subsets B_i the *blocks* of π and say that π has k blocks, denoted $|\pi| = k$. The number of partitions of S with k blocks is the Stirling number $S(n, k)$ of the second kind, while the total number of partitions of S is the Bell number $B(n)$ (see, e.g., [10, Ch. V]). Let Π_n denote the set of all partitions of $[n] = \{1, 2, \ldots, n\}$.

More generally, if r is a fixed positive integer and if S is an n-element set, define an *r-partition of S* to be a set

$$\pi = \{(B_{11}, B_{12}, \ldots, B_{1r}), (B_{21}, B_{22}, \ldots, B_{2r}), \ldots, (B_{k1}, B_{k2}, \ldots, B_{kr})\}$$

satisfying:

(i) For each $j \in [r]$, the set $\pi_j = \{B_{1j}, B_{2j}, \ldots, B_{kj}\}$ forms a

GENERATING FUNCTIONS 131

partition of S into k blocks. Thus each B_{ij} is a nonvoid subset of S, the k sets $B_{1j}, \ldots, B_{kj}$ are pairwise disjoint, and $\bigcup_i B_{ij} = S$.

(ii) For fixed i, $|B_{i1}| = |B_{i2}| = \cdots = |B_{ir}|$.

It follows that the r partitions $\pi_1, \ldots, \pi_r$ all have the same type $(a_1, a_2, \ldots, a_n)$, which we call the *type* of π. We also say that π has k blocks, denoted $|\pi| = k$. We let Π_{nr} denote the set of all r-partitions of $[n]$, so $\Pi_{n1} = \Pi_n$.

6.1. Theorem (the convolutional formula for r-partitions): *Let* $f: \mathbf{P} \to \mathbf{C}$ *and* $g: \mathbf{P} \to \mathbf{C}$. *Define a new function* $h: \mathbf{P} \to \mathbf{C}$ *by*

$$h(n) = \sum_\pi f(1)^{a_1} f(2)^{a_2} \cdots f(n)^{a_n} g(|\pi|),$$

where π *ranges over all r-partitions of $[n]$, and where π has type $(a_1, a_2, \ldots, a_n)$. Define the power series* $F(X), G(X), H(X) \in \mathbf{C}[[X]]$ *by*

$$F(X) = \sum_1^\infty f(n) X^n/n!^r, \quad G(X) = \sum_1^\infty g(n) X^n/n!,$$

$$H(X) = \sum_1^\infty h(n) X^n/n!^r.$$

Then $H(X) = G(F(X))$.

Proof: We have

$$G(F(X)) = \sum_{k=1}^\infty g(k) \left[\sum_{i=1}^\infty f(i) X^i/i!^r \right]^k / k!$$

$$= \sum_{k=1}^\infty \frac{g(k)}{k!} \sum \frac{f(b_1) f(b_2) \cdots f(b_k)}{b_1!^r b_2!^r \cdots b_k!^r} X^{b_1 + b_2 + \cdots + b_k},$$

where the inner sum is over all k-tuples $(b_1, \ldots, b_k) \in \mathbf{P}^k$. Let a_i be the number of b_j's which are equal to i, so that $k = \Sigma a_i$; and let $n = \Sigma b_i = \Sigma i a_i$. We obtain

$$G(F(X)) = \sum_{n=1}^{\infty} \frac{X^n}{n!^r} \sum \frac{n!^r \alpha(a_1, \ldots, a_n)}{(1!^{a_1} 2!^{a_2} \cdots n!^{a_n})^r k!} f(1)^{a_1} \cdots f(n)^{a_n} g(k),$$

where the inner sum is over all solutions in non-negative integers a_i to $n = \Sigma i a_i$, where $k = \Sigma a_i$, and where $\alpha(a_1, \ldots, a_n)$ is the number of distinct k-tuples $(b_1, \ldots, b_k)$ with exactly a_i of the b_j's equal to i. Clearly $\alpha(a_1, \ldots, a_n)$ is the multinomial coefficient $k!/a_1! a_2! \cdots a_n!$. Hence

$G(F(X)) =$

$$\sum_{n=1}^{\infty} \frac{X^n}{n!^r} \sum \frac{n!^r}{(1!^{a_1} \cdots n!^{a_n})^r a_1! \cdots a_n!} f(1)^{a_1} \cdots f(n)^{a_n} g(k).$$

It is easily proved that the number of r-partitions of $[n]$ of type $(a_1, \ldots, a_n)$ is just $n!^r/(1!^{a_1} \cdots n!^{a_n})^r a_1! \cdots a_n!$. From this the proof follows. □

6.2. Corollary (the exponential formula for r-partitions): *In Theorem 6.1, let $g(n) = 1$ for all $n \in \mathbf{P}$. Then*

$$1 + H(X) = \exp F(X). \square$$

More sophisticated approaches to Theorem 6.1 and Corollary 6.2 are given in [15, Ch. III], [14], [11, Thm. 5.1], and [7, §3]. The first three of these references treat only the case $r = 1$, and our viewpoint most closely follows [11]. The prefab theory of [7] gives more general results than Corollary 6.2, though it is possible to extend Theorem 6.1, in a manner analogous to our treatment of binomial posets in Section 3, so that it achieves the same level of generality as the treatment in [7].

We conclude this section with a number of applications of Theorem 6.1 and Corollary 6.2.

6.3. Example: If we set $f(n) = Y$ for all $n \in \mathbf{P}$ in Corollary 6.2, then $h(n) = \sum_k S_r(n, k) Y^k$, where $S_r(n, k)$ is the number of r-partitions of $[n]$ into k blocks. Hence from Corollary 6.2 we obtain

GENERATING FUNCTIONS

$$1 + \sum_{n=1}^{\infty} \sum_{k=1}^{\infty} S_r(n, k) X^n Y^k/n!^r = \exp Y \sum_{n=1}^{\infty} X^n/n!^r.$$

In particular (putting $Y = 1$),

$$1 + \sum_{1}^{\infty} |\Pi_{nr}| \cdot X^n/n!^r = \exp \sum_{n=1}^{\infty} X^n/n!^r.$$

When $r = 1$, $S_r(n, k)$ becomes the Stirling number $S(n, k)$ of the second kind, and $|\Pi_{nr}|$ becomes the Bell number $B(n)$. We recover the well-known results (see, e.g., [10, Ch. V])

$$1 + \sum_{n=1}^{\infty} \sum_{k=1}^{\infty} S(n, k) X^n Y^k/n! = \exp Y(e^X - 1),$$

$$1 + \sum_{n=1}^{\infty} B(n) X^n/n! = \exp(e^X - 1).$$

6.4. Example: Let $f(n)$ be the number of connected graphs (without loops or multiple edges) on the vertex set $[n]$, and let $h(n, k)$ be the total number of graphs on $[n]$ with k connected components. A graph with k components can be obtained uniquely by partitioning $[n]$ into k blocks and "attaching" a connected graph to each block. If a block B has i elements, then there are $f(i)$ connected graphs which can be placed on B. Hence

$$\sum_k h(n, k) Y^k = \sum_{\pi \in \Pi_n} f(1)^{a_1} \cdots f(n)^{a_n} Y^{a_1 + \cdots + a_n}$$

where in the latter sum π has type $(a_1, a_2, \ldots, a_n)$. From Corollary 6.2 (with $r = 1$) we obtain

$$1 + \sum_{n=1}^{\infty} \sum_{k=1}^{\infty} h(n, k) X^n Y^k/n! = \exp Y \cdot \sum_{1}^{\infty} f(n) X^n/n!.$$

Clearly there are a total of $2^{\binom{n}{2}}$ graphs on the vertex set $[n]$. Hence setting $Y = 1$ in the above formula, we get

$$\sum_{n=0}^{\infty} 2^{\binom{n}{2}} X^n/n! = \exp \sum_{n=1}^{\infty} f(n) X^n/n!. \tag{18}$$

Note that the above power series has zero radius of convergence, but this need not be a cause of concern since (18) is a *formal* power series identity.

Example 6.4 is the archetypal application of the exponential formula in the case $r = 1$. Whenever we have some structure on $[n]$ which is "pieced together" from its connected components, we obtain a formula analogous to (18). For instance, instead of graphs we could equally well have used partial orders, topologies, diagraphs, etc.

6.5. Example: Let $h(n)$ be the number of graphs G on the vertex set $[n]$, such that every component of G is a cycle (of length ≥ 3), an edge, or a single vertex. There are $(i - 1)!/2$ cycles on an i-element set, $i \geq 3$. (Of course we mean an undirected cycle; there are $(i - 1)!$ directed cycles.) Moreover, there is one two-vertex graph with an edge, and only one single-vertex graph. Hence from Corollary 6.2 (with $r = 1$) we get

$$1 + \sum_{1}^{\infty} h(n) X^n/n! = \exp\left[X + \frac{X^2}{2} + \frac{1}{2} \sum_{3}^{\infty} \frac{X^i}{i}\right]$$

$$= \exp\left[\frac{X}{2} + \frac{X^2}{4} - \frac{1}{2} \log(1 - X)\right]$$

$$= (1 - X)^{-1/2} \exp\left(\frac{X}{2} + \frac{X^2}{4}\right).$$

The function $h(n)$ has several other interesting combinatorial interpretations, e.g., (a) the number of distinct matrices of the form $P + P^{-1}$, where P is an $n \times n$ permutation matrix, and (b) the number of distinct monomials appearing in the expansion of the determinant of an $n \times n$ symmetric matrix whose entries x_{ij} are independent indeterminates (except $x_{ij} = x_{ji}$). For a modification of this result, see [10, Ch. 7.3].

6.6. Example: Suppose we have a room of n children. The children gather into circles by holding hands, and one child stands in the center of each circle. A circle may consist of as little as one child (clasping his or her hands), but each circle must contain a child inside it. In how many ways can this be done? Let this number be $h(n)$. An allowed arrangement of children is obtained by choosing a partition of the children, choosing a child c from each block B to be in the center of a circle, and arranging the other children in the block B in a circle about c. If $|B| = i \geq 2$, then there are $i \cdot (i - 2)!$ ways to do this, while there are no ways if $i = 1$. Hence

$$1 + \sum_{1}^{\infty} h(n) X^n/n! = \exp \sum_{2}^{\infty} \frac{i \cdot (i - 2)! X^i}{i!}$$

$$= \exp X \sum_{i=1}^{\infty} \frac{X^i}{i} = (1 - X)^{-X}.$$

The astute reader has undoubtedly realized by now that this example was contrived solely to obtain the curious answer. With a little practice such generating functions can be quickly computed in one's head.

6.7. Example: Let $\mathfrak{S}_n$ denote the group of all permutations of $[n]$. For fixed $m \in \mathbf{P}$, let $h(n)$ denote the number of $\rho \in \mathfrak{S}_n$ satisfying $\rho^m = 1$. Such a ρ can be obtained by partitioning $[n]$ into blocks B whose cardinality d divides m, and then choosing a cyclic permutation of B. Such a cyclic permutation can be chosen in $(d - 1)!$ ways. Hence

$$1 + \sum_{n=1}^{\infty} h(n) X^n/n! = \exp \sum_{d \mid m} \frac{(d - 1)! X^d}{d!}$$

$$= \exp \sum_{d \mid m} \frac{X^d}{d}.$$

More generally, the coefficient of $Y_1^{a_1} Y_2^{a_2} \cdots Y_n^{a_n} X^n/n!$ in exp $\sum_{i=1}^{\infty} Y_i X^i/i$ is equal to the number of $\rho \in \mathfrak{S}_n$ with exactly a_i cycles of length i. This well-known result (e.g., [33]) is equivalent to Corollary 6.2 in the case $r = 1$.

6.8. Example: Let $t(n)$ be the number of rooted trees (connected acyclic graphs with a distinguished vertex) on the vertex set $[n]$, and let $f(n)$ be the number of rooted forests (graphs whose components are rooted trees) on $[n]$. The reader who has come this far will instantaneously see that

$$1 + \sum_1^{\infty} f(n) X^n/n! = \exp \sum_1^{\infty} t(n) X^n/n!. \tag{19}$$

On the other hand, any tree on $[n + 1]$ gives rise to a rooted forest on $[n]$ by removing the vertex $n + 1$ and all incident edges, and putting roots at the vertices adjacent to $n + 1$. Since there are $n + 1$ ways of rooting a tree on $[n + 1]$, we get $(n + 1)f(n) = t(n + 1)$. Setting $T(X) = \sum_{n=1}^{\infty} t(n) X^n/n!$, equation (19) results in the famous functional equation $T(X) = X \cdot \exp T(X)$. Using the Lagrange inversion formula (Theorem 5.6), one easily deduces that $t(n) = n^{n-1}$. For further information on this result, including direct combinatorial proofs, see [29].

6.9. Example: Let $h(n)$ be the number of idempotent functions $\beta:[n] \to [n]$, i.e., $\beta(\beta(i)) = \beta(i)$ for all $i \in [n]$. A function $\beta:[n] \to [n]$ is idempotent if and only if for each $i \in [n]$, the set $\beta^{-1}(i)$ is empty or contains i. Hence we obtain an idempotent function by partitioning $[n]$ and mapping each element of a block B to a fixed element x of that block. If $|B| = i$ then there are i choices for x. Thus

$$1 + \sum_1^{\infty} h(n) X^n/n! = \exp \sum_1^{\infty} i X^i/i!$$
$$= \exp X e^X. \tag{20}$$

GENERATING FUNCTIONS 137

For further information on $h(n)$, see [21]. The reader may find it interesting to generalize (20) in various ways. For instance, given $1 \le i < j$, how many functions $\beta:[n] \to [n]$ satisfy $\beta^i = \beta^j$?

6.10. Example: Fix $s > 0$. Let $f_s(n)$ be the number of sequences $\rho_1, \rho_2, \ldots, \rho_s$ of s permutations in the symmetric group $\mathfrak{S}_n$ on $[n]$ which generate a transitive subgroup of $\mathfrak{S}_n$. There are $n!^s$ sequences $\rho_1, \rho_2, \ldots, \rho_s$ with no assumptions on transitivity. Given any such sequence, the orbits of the group generated by $\rho_1, \rho_2, \ldots, \rho_s$ form a partition of $[n]$. Given a partition π of $[n]$ of type $(a_1, a_2, \ldots, a_n)$, the number of sequences $\rho_1, \rho_2, \ldots, \rho_s$ with orbit partition π is clearly $f_s(1)^{a_1} f_s(2)^{a_2} \cdots f_s(n)^{a_n}$. Hence

$$n!^s = \sum_{\pi \in \Pi_n} f_s(1)^{a_1} \cdots f_s(n)^{a_n},$$

so by Corollary 6.2 (with $r = 1$),

$$1 + \sum_1^\infty n!^s X^n/n! = \sum_0^\infty n!^{s-1} X^n = \exp \sum_1^\infty f_s(n) X^n/n!$$

Now let F_s be the free group on generators $x_1, x_2, \ldots, x_s$. Let G be a subgroup of F_s of index n. Let $G_2, G_3, \ldots, G_n$ be an ordering ι (out of the $(n-1)!$ possible orderings) of the cosets of G not equal to G. Let $G = G_1$. Define permutations $\rho_1, \rho_2, \ldots, \rho_s$ in $\mathfrak{S}_n$ by $x_i G_j = G_{\rho_i(j)}$. It is easily seen that $\rho_1, \rho_2, \ldots, \rho_s$ generate a transitive subgroup of $\mathfrak{S}_n$. It follows from the theory of free groups (e.g., [19, Theorem 7.2.7]) that the map $(\iota, G) \to (\rho_1, \rho_2, \ldots, \rho_s)$ is a bijection between (a) pairs (ι, G), where G is a subgroup of F_s of index n and ι is an ordering of the $n - 1$ proper cosets of G, and (b) sequences $(\rho_1, \ldots, \rho_s) \in \mathfrak{S}_n^s$ whose elements generate a transitive subgroup of $\mathfrak{S}_n$. Thus if $N_s(n)$ denotes the number of subgroups of F_s of index n, then $N_s(n) = f_s(n)/(n-1)!$ and

$$\sum_{n=1}^\infty n!^{s-1} X^n = \exp \sum_{n=1}^\infty N_s(n) X^n/n. \tag{21}$$

A recursion equivalent to (21) appears in [19, Theorem 7.2.9].

From (21) E. Bender [6, §5] has derived an asymptotic expansion for $N_s(n)$ for fixed s.

6.11. Example: Let $n \in \mathbf{P}$ and $s \in \mathbf{N}$, and let $\mathfrak{M}(n, s)$ denote the set of all $n \times n$ matrices of nonnegative integers for which every row and column sums to s. Let $M \in \mathfrak{M}(n, s)$. We regard the rows and columns of M as being indexed by $[n]$; i.e., $M = (m_{ij})$, where $(i, j) \in [n] \times [n]$. By a k-*component* of M, we mean a pair (A, B) of subsets of $[n]$ satisfying the following two properties:

(i) $|A| = |B| = k \geq 1$,

(ii) Let $M(A, B)$ be the $k \times k$ submatrix of M whose rows are indexed by A and whose columns are indexed by B, i.e., $M(A, B) = (m_{ij})$, where $(i, j) \in A \times B$. Then every row and column of $M(A, B)$ sums to s, i.e., $M(A, B) \in \mathfrak{M}(k, s)$.

A component (A, B) is *irreducible* if any component (A', B') with $A' \subset A$ and $B' \subset B$ satisfies $(A', B') = (A, B)$. For instance, $(\{i\}, \{j\})$ is a 1-component (in which case it is irreducible) if and only if $m_{ij} = s$. It is easily seen that the irreducible components of M form a 2-partition of $[n]$. Conversely, any matrix $M \in \mathfrak{M}(n, s)$ can be obtained by choosing a 2-partition π of $[n]$ and "attaching" an irreducible component to each $(A, B) \in \pi$. Let $h_s(a_1, \ldots, a_n)$ denote the number of matrices $M \in \mathfrak{M}(n, s)$ such that M has a_i irreducible i-components (so $n = \Sigma i a_i$). Let $f_s(n)$ be the number of irreducible $n \times n$ matrices $M \in \mathfrak{M}(n, s)$, i.e., $([n], [n])$ is an irreducible component of M. It then follows from Corollary 6.2 that

$$\sum_{n=0}^{\infty} \sum_{a_1, \ldots, a_n} h_s(a_1, \ldots, a_n) Y_1^{a_1} \cdots Y_n^{a_n} X^n / n!^2 \qquad (22)$$

$$= \exp \sum_{1}^{\infty} f_s(n) Y_n X^n / n!^2.$$

In particular, if $H(n, s) = |\mathfrak{M}(n, s)|$, then

$$\sum_{0}^{\infty} H(n, s) X^n / n!^2 = \exp \sum_{1}^{\infty} f_s(n) X^n / n!^2.$$

Similarly, let $H^*(n, s)$ denote the number of matrices in $\mathfrak{M}(n, s)$ with no entry equal to s. Since $f_s(1) = 1$, there follows

$$\sum_0^\infty H^*(n, s)X^n/n!^2 = \exp \sum_2^\infty f_s(n)X^n/n!^2$$

$$= e^{-X} \sum_0^\infty H(n, s)X^n/n!^2.$$

It is not difficult to compute $f_2(n)$. Indeed, an irreducible matrix $M \in \mathfrak{M}(n, 2)$ is of the form $P + PQ$, where P is a permutation matrix and Q is a cyclic permutation matrix. There are $n!$ choices for P and $(n - 1)!$ choices for Q. If $n > 1$ then P and PQ could have been chosen in reverse order. Hence $f_2(1) = 1$ and $f_2(n) = n!(n - 1)!/2$ if $n > 1$. There follows

$$\sum_{n=0}^\infty \sum_{a_1,\ldots,a_n} h_2(a_1, \ldots, a_n)Y_1^{a_1} \cdots Y_n^{a_n}X^n/n!^2$$

$$= \exp\left[\frac{Y_1 X}{2} + \frac{1}{2}\sum_{n=1}^\infty \frac{Y_n X^n}{n}\right].$$

In particular,

$$\sum_0^\infty H(n, 2)X^n/n!^2 = (1 - X)^{-1/2} e^{X/2}, \tag{23}$$

$$\sum_0^\infty H^*(n, 2)X^n/n!^2 = (1 - X)^{-1/2} e^{-X/2}, \tag{24}$$

Equations (23) and (24) are due to Anand, Dumir, and Gupta [3]. For additional information about the functions $H(n, s)$, see [40], [10, pp. 124-125]. It appears certain, however, that there are no formulas for $H(n, 3)$ as simple as (23). The reader may find it of interest to derive a formula analogous to (23) involving $S(n, 2)$, the number of *symmetric* matrices in $\mathfrak{M}(n, 2)$ (see [18]).

REFERENCES

1. H. L. Alder, "Partition identities—from Euler to the present", *Amer. Math. Monthly*, **76** (1969), 733-746.
2. R. Alter, "The Catalan Numbers", 1971 Proc. Louisiana Conference on Combinatorics, Graph Theory and Computing.
3. H. Anand, V. C. Dumir, and H. Gupta, "A combinatorial distribution problem", *Duke Math. J.*, **33** (1966), 757-769.
4. G. E. Andrews, *Number Theory*, Saunders, Philadelphia, 1971.
5. ___, "Partition identities", *Advances in Math.*, **9** (1972), 10-51.
6. E. A. Bender, "Asymptotic methods in enumeration", *SIAM Rev.*, **16** (1974), 485-515.
7. E. A. Bender and J. R. Goldman, "Enumerative uses of generating functions", *Indiana Univ. Math. J.*, **20** (1971), 753-765.
8. W. G. Brown, "Historical note on a recurrent combinatorial problem", *Amer. Math. Monthly*, **72** (1965), 973-977.
9. L. Comtet, "Calcul pratique des coefficients de Taylor d'une fonction algébrique", *Enseignement Math.*, **10** (1964), 267-270.
10. ___, *Advanced Combinatorics*, Reidel, Dordrecht and Boston, 1974.
11. P. Doubilet, G.-C. Rota, and R. Stanley, "On the Foundations of Combinatorial Theory (VI): The Idea of Generating Function", Sixth Berkeley Symposium on Mathematical Statistics and Probability, Vol. II: Probability Theory, University of California 1972, 267-318.
12. E. Ehrhart, "Sur la loi de réciprocité des polyèdres rationnels", *C. R. Acad. Sci. Paris*, **266A** (1968), 696-697.
13. I. M. H. Etherington and A. Erdélyi, "Some problems of non-associative combinations, II", *Edinburgh Math. Notes*, **32** (1940), 7-12.
14. D. Foata, *La Série Génératrice Exponentielle dans les Problèmes d'Énumération*, Les Presses de l'Université de Montréal, 1974.
15. D. Foata and M.-P. Schützenberger, "Théorie Géométrique des Polynômes Eulériens", *Lecture Notes in Math.*, 138, Springer-Verlag, Berlin, 1970.
16. J. Goldman and G.-C. Rota, "On the foundations of combinatorial theory IV: Finite vector spaces and Eulerian generating functions", *Studies in Appl. Math.*, **49** (1970), 239-258.
17. H. W. Gould, "Bell and Catalan numbers", published by the author, 1976; revised and corrected, 1977.
18. H. Gupta, "Enumeration of symmetric matrices", *Duke Math. J.*, **35** (1968), 653-659.
19. M. Hall, Jr., *The Theory of Groups*, Macmillan, New York, 1959.
20. G. H. Hardy and E. M. Wright, *An Introduction to the Theory of Numbers*, 4th ed., Oxford University Press, 1960.
21. B. Harris and L. Schoenfeld, "The number of idempotent elements in symmetric semigroups", *J. Combinatorial Theory*, **3** (1967), 122-135.
22. M. Henle, "Dissection of generating functions", *Studies in Appl. Math.*, **51** (1972), 397-410.
23. D. A. Klarner, "Cell growth problems", *Canad. J. Math.*, **19** (1967), 851-863.

24. ____, "A combinatorial formula involving the Fredholm integral equation", *J. Combinatorial Theory*, **5** (1968), 59-74.
25. D. A. Klarner and M. L. J. Hautus, "The diagonal of a double power series", *Duke Math. J.*, **38** (1971), 229-235.
26. D. E. Knuth, *The Art of Computer Programming*, 3 volumes, Addison-Wesley, Reading, Mass., vol. 1 (1968, 2nd ed. 1973), vol. 2 (1969), vol. 3 (1973).
27. P. A. MacMahon, "The indices of permutations....", *Amer. J. Math.*, **35** (1913), 281-322.
28. ____, *Combinatory Analysis*, vols. 1-2, Cambridge University Press, 1916; repr. by Chelsea, New York, 1960.
29. J. W. Moon, "Counting Labelled Trees", Canadian Mathematical Monographs, No. 1, Canadian Math. Congress, 1970.
30. T. Popoviciu, "Studie si cercetari stiintifice", *Acad. R. P. R., Filiala Cluj*, **4** (1953), 8.
31. G. N. Raney, "Functional composition patterns and power series reversion", *Trans. Amer. Math. Soc.*, **94** (1960), 441-451.
32. R. Read, "The number of k-colored graphs on labelled nodes", *Canad. J. Math.*, **12** (1960), 410-414.
33. J. Riordan, *An Introduction to Combinatorial Analysis*, Wiley, New York, 1958.
34. G.-C. Rota, "On the foundations of combinatorial theory, I. Theory of Möbius functions", *Z. Wahrscheinlichkeitstheorie und Verw. Gebiete*, **2** (1964), 340-368.
35. E. Schröder, "Vier Combinatorische Probleme", *Z. Math. Phys.*, **15** (1870), 361-376.
36. R. Stanley, "Theory and application of plane partitions, parts 1 and 2", *Studies in Appl. Math.*, **50** (1971), 167-188, 259-279.
37. ____, "Ordered structures and partitions", *Mem. Amer. Math. Soc.*, **119** (1972).
38. ____, "Supersolvable lattices", *Algebra Universalis*, **2** (1972), 197-217.
39. ____, "Acyclic orientations of graphs", *Discrete Math.*, **5** (1973), 171-178.
40. ____, "Linear homogeneous diophantine equations and magic labelings of graphs", *Duke Math. J.*, **40** (1973), 607-632.
41. ____, "Combinatorial reciprocity theorems", *Advances in Math.*, **14** (1974), 194-253.
42. D. Tamari, "The algebra of bracketings and their enumeration", *Nieuw. Arch. Wisk.*, (3) **10** (1962), 131-146.
43. E. T. Whittaker and G. N. Watson, *A Course in Modern Analysis*, 4th ed., Cambridge University Press, 1927.
44. G. Andrews, "The Theory of Partitions", *Encyclopedia of Mathematics and Its Applications* (G.-C. Rota, ed.), vol. 2, Addison-Wesley, Reading, Mass., 1976.
45. I. Gessel and R. Stanley, "Stirling polynomials", *J. Combinatorial Theory*, to appear.
46. G. Pólya, "On the number of certain lattice polygons", *J. Combinatorial Theory*, **6** (1969), 102-105.

NONCONSTRUCTIVE METHODS IN DISCRETE MATHEMATICS

Joel Spencer

0. INTRODUCTION

Discrete mathematics covers a broad spectrum of problems intersecting nearly all branches of modern mathematics. One fundamental problem is the demonstration of the existence of finite configurations satisfying certain prescribed properties. Here we consider a *method* which proves the existence of such configurations without actually constructing them. We call this method the *nonconstructive* or *probabilistic* method.

We can enumerate a triad of possible techniques to prove the existence of finite configurations. *Direct* methods (i.e., actual constructions) are the most common. *Recursive*, or *inductive*, methods are also well known. Nonconstructive methods are considerably less well known. They have only recently received a wide audience.

The first use of this method is credited to Szele [16]. However, it was the application of nonconstructive methods to obtain a lower bound on Ramsey Numbers (see Section 1) that truly began the development of this area. Development of the probabilistic method of proof is largely due to the work of Paul Erdös. A more detailed account of the probabilistic method can be found in Probabilistic Methods in Combinatorics by Paul Erdös and Joel Spencer [8].

We have applied the probabilistic method to a variety of discrete problems to afford a more pragmatic examination of the method itself. These appear in a series of self-contained sections.

Section 1 considers the lower bound to the Ramsey Number $R(k, k)$. The proof techniques, which are discussed in fair detail, give a functional definition of nonconstructive methods. Section 2 presents a variety of tournament problems and conjectures. Section 3 gives Paul Erdös' aesthetically pleasing result on the existence of graphs with arbitrarily high chromatic number and girth. In Section 4 we turn to a discussion of random graphs *per se* which requires an interface of the concepts of graph theory and probability theory. Such graphs are used by both social and natural scientists in various models of natural phenomena. Section 5 deals with coding theory. We examine the celebrated result of Claude Shannon on the existence of codes with positive rate of transmission and arbitrarily low rate of error.

First, it may be helpful to standardize a small amount of notation:

$[n] = \{1, 2, \ldots, n\}$ will denote a canonical n-element set;

$[S]^k = \{T: T \subseteq S, |T| = k\}$;

$[n]^k = \{T: T \subseteq [n], |T| = k\}$.

A graph G consists of a vertex set $V = V(G)$ and an edge set $E = E(G) \subseteq [V]^2$. In standard notation, all graphs are assumed undirected, without loops or multiple edges.

We let P denote probability, E expected value, $\overline{A}$ denote the negative of A, $\wedge$ conjunction, and $\vee$ disjunction. We will assume a moderate acquaintance with the basic concepts of probability theory. In particular, we assume the linearity of expected value. That is $E(X + Y) = E(X) + E(Y)$ whether or not X and Y are independent.

We define

$$n! = \prod_{i=1}^{n} i,$$

$$\binom{n}{k} = n!/(k!(n-k)!),$$

$$(n)_k = \prod_{i=0}^{k-1} (n-i) = n!/(n-k)!.$$

We use $o(f(n))$ to denote a function $g(n)$ satisfying $\lim g(n)/f(n) = 0$. We write $O(f(n))$ to denote a function $g(n)$ satisfying $\limsup g(n)/f(n) < \infty$. In particular, $o(1)$ represents a function of n (the parameter) which approaches zero, while $O(1)$ represents a bounded function.

We will assume Stirling's Formula

$$n! = n^n e^{-n} \sqrt{2\pi n}\, (1 + o(1)).$$

1. RAMSEY'S THEOREM

In any collection of six people either some three mutually know each other or some three mutually do not know each other.

This result, found in numerous problem books and mathematical competitions, is the canonical example of Ramsey's Theorem. We define $R(k, l)$ as the minimal n such that in any collection of n people either some k mutually know each other or some l mutually do not. In this section we examine bounds on the function $R(k, l)$. A survey of applications and generalizations of Ramsey's Theorem is given by R. L. Graham and B. Rothschild elsewhere in this volume.

Let us reintroduce the function R in a more formal context. Notation: Let $G = (V, E)$ be a graph. A set $S \subseteq V$ is a *clique* if $[S]^2 \subseteq E$. A set $S \subseteq V$ is called an *independent set* if $[S]^2 \cap E = \emptyset$. The clique number of G, denoted by $\omega(G)$, is defined by

$$\omega(G) = \max\{|S|: S \subseteq V, S \text{ a clique}\}.$$

Similarly, the independence number, denoted by $\alpha(G)$, is defined by

$$\alpha(G) = \max\{|S|: S \subseteq V, S \text{ an independent set}\}.$$

DEFINITION: Let $k, l \geq 2$. The Ramsey Number $R(k, l)$ is the minimal integer n such that if G is a graph on n vertices, $\omega(G) \geq k$ or $\alpha(G) \geq l$.

We can correspond the "knowing relationship" among n people to a graph on n points by joining two vertices if and only if the corresponding people know each other. Our two definitions of R are then seen to be equivalent.

The existence of $R(k, l)$ for all k, l is given by Ramsey's Theorem. In fact, we can show

$$R(k, l) \leq \binom{k + l - 2}{k - 1}. \tag{1.1}$$

In this section we will be interested in *lower* bounds to $R(k, k)$. That is, we wish to find graphs G on many vertices with $\omega(G), \alpha(G) \leq k$.

THEOREM 1.1 (Erdös [4]): *If*

$$\binom{n}{k} 2^{\binom{n}{2} - \binom{k}{2} + 1} < 2^{\binom{n}{2}}, \tag{1.2}$$

then $R(k, k) > n$.

We can interpret Theorem 1.1 by the following corollary:

COROLLARY 1.1: $R(k, k) > k \, 2^{k/2} \left[\dfrac{1}{e\sqrt{2}} + o(1) \right].$

The corollary follows from an elementary application of Stirling's Formula. Corollary 1.1 and (1.1) bound $R(k, k)^{1/k}$ between $\sqrt{2} + o(1)$ and $4 + o(1)$. The value of $\lim R(k, k)^{1/k}$ is not known. The determination of this limit is a major open problem. Even the existence of the limit itself has not yet been proven.

Proof of Theorem 1.1: Fix n satisfying (1.1). There are $2^{\binom{n}{2}}$ graphs G on vertex set $[n]$. Call G *good* if $\omega(G), \alpha(G) \leq k$. Call G *bad* if it is not good. For any $S \in [n]^k$ there are exactly $2^{\binom{n}{2} - \binom{k}{2} + 1}$ graphs G in which S is a clique or an independent set. If G is *bad*, some $S \in [n]^k$ is a clique or independent set. Therefore, there are at most $\binom{n}{k} 2^{\binom{n}{2} - \binom{k}{2} + 1}$ G that are bad. Assuming (1.2), some G must be good.

We have set forth an example (perhaps *the* basic example) of the *nonconstructive* method of proof. We have shown the existence of a good G by demonstrating that there are not enough bad G's. Consequently, we are left with an uneasy feeling because no *specific* good G has been *constructed*. The situation may appear analogous to the Axiom of Choice, where existential questions rely on the fundamental premises of set theory. However, let us quickly point out that there are no logical difficulties in the nonconstructive method. We can always enumerate *all* G on $[n]$ until we discover the desired good graph.

In order to develop further our basic methodologies, we now re-prove Theorem 1.1 using the probabilistic approach.

THEOREM 1.1': *If*

$$\binom{n}{k} 2^{1 - \binom{k}{2}} < 1, \tag{1.3}$$

then $R(k, k) > n$.

Proof: Let **G** be a *random* graph on the vertex set $[n]$ where each edge $\{i, j\}$ is in **G** with probability $\frac{1}{2}$ and these probabilities are mutually independent. Technically, we are dealing with a sample space of $2^{\binom{n}{2}}$ points; one for each possible G, each with probability $2^{-\binom{n}{2}}$. For any $S \in [n]^k$ let A_S be the event "S is a clique or an independent set in **G**." Then

$$P(A_S) = 2^{1-\binom{k}{2}}. \tag{1.4}$$

Hence,

$$P(\vee A_S) \le \Sigma P(A_S)$$
$$= \binom{n}{k} 2^{1-\binom{k}{2}} \tag{1.5}$$

$$< 1, \quad \text{by assumption.}$$

Here the disjunction and summation are over all $S \in [n]^k$. Therefore,

$$P(\wedge \overline{A_S}) > 0. \tag{1.6}$$

But $\wedge \overline{A_S}$ is precisely the event "**G** is good." An event with positive probability cannot be the null set. Consequently, there must be a specific good G. Therefore $R(k, k) > n$.

In proving Theorem 1.1 we use a *counting argument*, whereas we prove Theorem 1.1′ by means of a *probabilistic* argument. We distinctly prefer the latter method. A probabilistic approach enables us to use theorems from probability that would be difficult to formulate in a counting argument. Another perhaps not so minor consideration is that the use of the probabilistic method generally results in a cleaner, more succinct, proof which is both easier to write and to comprehend.

We improve Theorem 1.1′ by taking advantage of the independence of A_S and A_T when $|S \cap T| \le 1$. First, we require a general theorem on probability.

THEOREM 1.2 (Lovász Local Theorem [6]): *Let A_i, $1 \le i \le n$, be events with $P(A_i) \le p$. Let G be a finite graph of maximal degree d on vertex set $[n]$ and assume A_i is jointly independent of $\{A_j : \{i, j\} \notin G\}$. If $4dp < 1$, then*

$$P\left(\bigwedge_{i=1}^{n} \overline{A_i}\right) > 0.$$

Proof: We show, by induction on m, that

$$P(A_1 | \overline{A_2} \ldots \overline{A_m}) \le \frac{1}{2d}. \tag{1.7}$$

Renumber so that A_1 is independent of $\{A_i : i > d + 1\}$. Then

$$P(A_1 | \overline{A_2} \ldots \overline{A_m}) = \frac{P(A_1 \overline{A_2} \ldots \overline{A_{d+1}} | \overline{A_{d+2}} \ldots \overline{A_m})}{P(\overline{A_2} \ldots \overline{A_{d+1}} | \overline{A_{d+2}} \ldots \overline{A_m})}. \tag{1.8}$$

We bound the numerator of (1.8) by

$$P(A_1 \overline{A_2} \ldots \overline{A_{d+1}} | \overline{A_{d+2}} \ldots \overline{A_m}) \le P(A_1 | \overline{A_{d+2}} \ldots \overline{A_m}) = P(A_1) \le p \tag{1.9}$$

since A_1 is independent of $\overline{A_{d+2}} \ldots \overline{A_m}$. We bound the denominator of (1.8) by

$$P(\overline{A_2} \ldots \overline{A_{d+1}} | \overline{A_{d+2}} \ldots \overline{A_m}) \ge 1 - \sum_{i=2}^{d-1} P(A_i | \overline{A_{d+2}} \ldots \overline{A_m})$$

$$\ge 1 - \sum_{i=2}^{d+1} \frac{1}{2d} \text{ (by induction)}$$

$$= \frac{1}{2} \tag{1.10}$$

Equations (1.9), (1.10) yield

$$P(A_1|\overline{A_2} \ldots \overline{A_m}) \le p/(\tfrac{1}{2}) = 2p < \tfrac{1}{2d} \qquad (1.11)$$

proving (1.7). Finally,

$$P\left(\bigwedge_{i=1}^{n} \overline{A_i}\right) = \prod_{i=1}^{n} P(\overline{A_i}|\overline{A_{i+1}} \ldots \overline{A_n})$$

$$\ge \prod_{i=1}^{n} \left(1 - \tfrac{1}{2d}\right)$$

$$= \left(1 - \tfrac{1}{2d}\right)^n > 0. \qquad (1.12)$$

THEOREM 1.3:

$$R(k, k) > k\, 2^{k/2}[\sqrt{2}/e + o(1)].$$

Proof: The event A_S is jointly independent of $\{A_T: |S \cap T| \le 1\}$ because these A_T involve different edges of **G** than A_S. We can apply the Lovász Local Theorem with

$$d = |\{T: |S \cap T| > 1\}| \le \binom{k}{2}\binom{n}{k-2}.$$

Thus, if

$$4\binom{k}{2}\binom{n}{k-2} 2^{1-\binom{k}{2}} < 1,$$

then

$$P(\wedge \overline{A_S}) > 0,$$

and therefore there exists a good G and $R(k, k) > n$. Stirling's Formula yields Theorem 1.3, an improvement of Corollary 1.1 by a factor of two.

We know that if $n \ge R(k, k)$, a random graph **G** has positive probability of being good. If we must actually construct a good graph, it is of interest to know $P[\mathbf{G}$ is good$]$. If this probability is near unity, we can, in any practical situation, choose G at random. From Theorem 1.1,

$$P[\mathbf{G} \text{ is bad}] \leq \binom{n}{k} 2^{1-\binom{k}{2}}. \tag{1.13}$$

For example, let $k = 100$. Then

$$P(\mathbf{G} \text{ is bad}) \leq n^{100} 2^{-4949}(100!)^{-1}$$

$$\sim n^{100} 10^{-1548}. \tag{1.14}$$

Thus, if $n = 3 \cdot 10^{15}$, $P[\mathbf{G}$ is bad$] < 1$ and therefore $R(100,100) < n$. By decreasing n slightly, say to $n = 2.5 \cdot 10^{15}$, we find

$$P[\mathbf{G} \text{ is bad}] \leq 10^{-9} \tag{1.15}$$

and now "almost all" G on n vertices are good.

We note that Theorem 1.3 gives $P[\mathbf{G}$ is bad$] < 1$ for $n \sim 6 \cdot 10^{15}$. However, for n "slightly less" it is no longer true that most G are good. Lovász's method yields an n for which good G exist but for which a random $\mathbf{G}$ is almost always bad.

Theorem 1.1′ can be generalized to give lower bounds to $R(k, l)$, $k \neq l$. Let $\mathbf{G} = \mathbf{G}_{n,p}$ be a random graph on n points such that $P[\{i, j\} \in \mathbf{G}] = p$ and these probabilities are mutually independent. For $S \in [n]^k$ let A_S be the event "S is a clique" and for $T \in [n]^l$ let B_T be the event "T is an independent set." Then

$$P[\vee A_S \vee \vee B_T] \leq \binom{n}{k} p^{\binom{k}{2}} + \binom{n}{l} (1-p)^{\binom{l}{2}} \tag{1.16}$$

from which we deduce

THEOREM 1.4: *If there exists p, $0 \leq p \leq 1$, such that*

$$\binom{n}{k} p^{\binom{k}{2}} + \binom{n}{l}(1-p)^{\binom{l}{2}} < 1, \qquad (1.17)$$

then $R(k, l) > n$.

2. TOURNAMENTS

In this section we study the properties of round robin tournaments. These tournaments consist of $\binom{n}{2}$ games among n players, each pair of players playing once, without draws. We will examine some problems connected with ranking the players in a tournament and the existence of tournaments having certain prescribed properties. The *random* tournament, whose games are decided by independent fair coin flips, will be employed to derive our results. The basic source book on this subject is J. W. Moon's *Topics on Tournaments* [13].

We now require some formal definitions. A tournament T_n on n players is a directed graph on a vertex set $[n]$ (each vertex is a "player") such that for $x, y \in V$, $(x, y) \in T$, or $(y, x) \in T$, but not both. The statement $(x, y) \in T$ represents "x beat y". We will let $\mathbf{T} = \mathbf{T}_n$ denote a random tournament on n players. For each i, j $P[(i, j) \in \mathbf{T}] = \frac{1}{2}$ *and these probabilities are independent.*

Schütte defined a tournament T to have property $S(k)$ if every k players $x_1, \ldots, x_k$ are beaten by some player y. He asked whether or not such T existed for every k. This is surprisingly difficult to show by direct methods (see Graham, Spencer [10]). However, when probabilistic means are employed, it becomes elementary.

Heretowit: For $W \in [n]^k$, $y \in [n] - W$, let A_{Wy} denote the event "y beat all $x \in W$" and B_W denote

$$\bigwedge_y \overline{A_{Wy}} = \text{"no } y \in [n] - W \text{ beat all } x \in W\text{"}.$$

Then

$$P[A_{Wy}] = 2^{-k} \tag{2.1}$$

and

$$P[B_W] = (1 - 2^{-k})^n, \tag{2.2}$$

since the A_{Wy} are independent over $y \in [n] - W$. Note that the event "not $S(k)$" is precisely $V B_W$; the disjunction over all $W \in [n]^k$. Thus,

$$P[\text{not } S(k)] \leq \binom{n}{k} (1 - 2^{-k})^n \tag{2.3}$$

yielding

THEOREM 2.1: *If*

$$\binom{n}{k}(1 - 2^{-k})^n < 1, \tag{2.4}$$

there exists T_n with property $S(k)$.

By elementary approximations, (2.4) holds if

$$n > 2^k k^2 (\ln 2 + o(1)). \tag{2.5}$$

We now turn our attention to the basic problem of ranking the players in a tournament. A ranking is a permutation σ of the vertex set $[n]$. There exist several differing criteria for determining the "best" ranking of a tournament T. Here we restrict our attention to one such criterion. For a given T let the *fit* of a ranking σ, denoted by $f(T, \sigma)$, be the number of $\{i, j\}$ such that $(i, j) \in T$ and $\sigma i < \sigma j$ (that is, the ranking agreed with the actual game). The "best" ranking is that σ which maximizes $f(T, \sigma)$. Of course, there may be many such σ. Call

$$g(T) = \max_{\sigma} f(T, \sigma) \tag{2.6}$$

the *consistency* of T. If T is transitive (i.e., $(i, j), (j, k) \in T$ imply $(i, k) \in T$) then there exists σ so that $f(T, \sigma) = \binom{n}{2}$. Thus $g(T) = \binom{n}{2}$. On the other hand, for any T if σ is any ranking

$$f(T, \sigma) + f(T, \bar{\sigma}) = \binom{n}{2},$$

where $\bar{\sigma}$ is the reverse ranking of σ. Hence $g(T) \geq \frac{1}{2}\binom{n}{2}$. Define

$$h(n) = \min g(T_n), \tag{2.7}$$

the minimum taken over all tournaments T_n on $[n]$.

THEOREM 2.2 (Erdös, Moon [7]):

$$h(n) \leq \tfrac{1}{2}\binom{n}{2} + \tfrac{1}{2}n^{3/2}\sqrt{\ln n}.$$

In particular, $h(n) \leq \frac{1}{2}\binom{n}{2}(1 + o(1))$. That is, for any $\epsilon > 0$, if n is sufficiently large there are tournaments T for which no ranking fits $\frac{1}{2} + \epsilon$ of the games.

No *direct* proof of Theorem 2.2 is known. Perhaps, in some sense that is not currently understood, one could show that it is *impossible* to find a direct construction. It may be that the very *concept* of a direct construction must, of necessity, place such order on T that some ranking will have a "good" fit. For the moment we have only random speculations. A precise definition of a "direct construction", which might involve notions of computational complexity, would be required before precise conjectures could even be stated, much less proved.

Proof of Theorem 2.2: Let $\mathbf{T}$ be a random tournament. For any fixed σ, $f(\mathbf{T}, \sigma)$ has distribution $\mathbf{U} = \mathbf{U}_{\binom{n}{2}}$, the sum of $\binom{n}{2}$ independent variables, each equiprobably 0 or 1. Now $\mathbf{U}$ is approximately a normal distribution with mean $\frac{1}{2}\binom{n}{2}$ and standard deviation $\left[\binom{n}{2}/4\right]^{\frac{1}{2}} \sim n/2\sqrt{2}$. Consequently **U** is rarely much greater than $\frac{1}{2}\binom{n}{2}$. To prove Theorem 2.2 we require information on the extreme tail of the distribution **U**. We can show [2]:

$$P\left[\mathbf{U} > \tfrac{1}{2}\binom{n}{2} + \alpha\right] < e^{-4\alpha^2/n^2}. \quad (2.8)$$

Setting $\alpha = \frac{1}{2}n^{3/2}\sqrt{\ln n}$, we have

$$e^{-4\alpha^2/n^2} = n^{-n},$$

and, consequently

$$P\left[f(\mathbf{T}, \sigma) > \tfrac{1}{2}\binom{n}{2} + \alpha\right] < n^{-n} < (n!)^{-1} \quad (2.9)$$

for any particular σ. Thus

$$P\left[g(\mathbf{T}) > \tfrac{1}{2}\binom{n}{2} + \alpha\right] = P\left[f(\mathbf{T}, \sigma) > \tfrac{1}{2}\binom{n}{2} + \alpha \text{ for some } \sigma\right] \quad (2.10)$$

$$\leq \sum_\sigma P\left[f(\mathbf{T}, \sigma) > \tfrac{1}{2}\binom{n}{2} + \alpha\right]$$

$$< n! n!^{-1} = 1.$$

Therefore, there exists T, $g(T) \leq \frac{1}{2}\binom{n}{2} + \alpha$, so $h(n) \leq \frac{1}{2}\binom{n}{2} + \alpha$.

Essentially, we have stated that $f(T, \sigma)$ is rarely very large and there are "only" $n!$ possible σ, and therefore $f(T, \sigma)$ is generally small for *all* σ's. Using what is normally considered the extreme tail of the binomial distribution curve is quite common in the probabilistic method. It is often necessary to obtain extremely small probabilities (e.g., $n!^{-1}$) to insure that they remain small when added over a large index set.

NONCONSTRUCTIVE METHODS IN DISCRETE MATHEMATICS 155

THEOREM 2.3. (Spencer [17]): $h(n) \geq \frac{1}{2}\binom{n}{2} + cn^{3/2}$, where c is a positive absolute constant.

Proof: Let us assume, for convenience, that n is odd, say $n = 2m + 1$. Fix T on player set $[n]$. We intend to select a $B \in [n]^m$ and to rank all $x \in [n] - B$ who beat more than half of B above B.

Define, for $x \in [n]$,

$$W_x = \{y \in [n] - \{x\}, (x, y) \in T\},$$

$$L_x = [n] - \{x\} - W_x = \{y \in [n] - \{x\}, (y, x) \in T\},$$

$$s_x = |W_x| = \text{score of } x.$$

For $B \in [n]^m$, $x \in [n] - B$, define

$$h(x, B) = |W_x \cap B| - |L_x \cap B| \qquad (2.11)$$

$$= 2|W_x \cap B| - m.$$

We wish to show that the "average" $|h(x, B)|$ is of the order $\sqrt{n}$.
Set

$$s = \sum_{x \in [n]} \sum_{\substack{B \in [n]^m \\ x \notin B}} |h(x, B)| = \sum_{B \in [n]^m} \sum_{x \notin B} |h(x, B)|. \qquad (2.12)$$

Fix $x \in [n]$ and assume $|W_x| = m$. If **B** is a random member of $[[n] - \{x\}]^m$, then $|W_x \cap \mathbf{B}|$ has a hypergeometric distribution which can be approximated by a normal distribution. We can approximate $|W_x \cap \mathbf{B}|$ by a normal distribution of mean $\mu = n/2$ and variance $\sigma^2 = (m/(2m - 1))(m/4) \sim n/16$. (This is the formula for sampling without replacement. See, e.g., Dwass [3, p. 322]). By (2.11), $h(x, \mathbf{B})$ is approximately a normal distribution with mean $\mu = 0$ and variance $\sigma^2 \sim n/4$. By elementary calculations

$$E[|h(x, \mathbf{B})|] \sim \sqrt{\frac{2}{\pi}} \cdot \sqrt{\frac{n}{4}} = \sqrt{n}/\sqrt{2\pi}. \qquad (2.13)$$

Thus

$$\sum_{\substack{B \in [n]^m \\ x \notin B}} |h(x, B)| \sim (\sqrt{n}/\sqrt{2\pi}) \cdot \binom{2m}{m}. \tag{2.14}$$

If $|W_x| \neq m$ then $h(x, B)$ has mean offcenter and it is relatively simple to show that the summation of (2.14) is even larger. The use of probabilistic means to prove (2.14) was not absolutely necessary—we could employ a straight counting argument. However, it can be argued that the probabilistic method gives more insight into the derivation of (2.14).

From (2.14) we deduce

$$s \geq n(\sqrt{n}/\sqrt{2\pi})\binom{2m}{m}, \tag{2.15}$$

and thus, by (2.12), for some $B \in [n]^m$

$$\sum_{x \notin B} |h(x, B)| \geq n^{3/2}(2\pi)^{-\frac{1}{2}}\binom{2m}{m}\binom{n}{m}^{-1}$$
$$\sim n^{3/2}/2\sqrt{2\pi}. \tag{2.16}$$

Fix this B. The summation of $|h(x, B)|$ may contain both positive and negative terms $h(x, B)$. Assume that the positive $h(x, B)$ contribute at least one half the sum (the negative case is similar). Set $X = \{x : h(x, B) \geq 0\}$. Then

$$\sum_{x \in X} h(x, B) \geq n^{3/2}/4\sqrt{2\pi}. \tag{2.17}$$

Now we can rank T. Rank B, X, and $[n] - X - B$ internally such that at least half the games agree with the ranking. Now rank $X < B$. That is, place every $x \in X$ above every $b \in B$. Of the games $\{x, b\}$, exactly

$$\sum_{x \in X} |W_x \cap B| = \sum_{x \in X} (h(x, B) + m)/2 \tag{2.18}$$

$$\geq |X||B|/2 + n^{3/2}/8\sqrt{2\pi}$$

agree with the ranking. Finally, rank $[n] - X \cup B$ with respect to $X \cup B$ such that at least half the games agree with the ranking. The final ranking σ has

$$f(T, \sigma) \geq \tfrac{1}{2}\binom{n}{2} + n^{3/2}/8\sqrt{2\pi}, \tag{2.19}$$

proving Theorem 2.3 for $c = 1/8\sqrt{2\pi}$.

Finding both bounds to $h(n)$ by probabilistic means is unusual. Furthermore, the proof of Theorem 2.3 is unusual in that it uses probabilistic means to find an asymmetry in *any* fixed T. **Added in proof:** Recently, this author has proven

$$h(n) < \tfrac{1}{2}\binom{n}{2} + c'n^{3/2}$$

using more complicated probabilistic methods. The proof will appear in *Periodica Math. Hung.*

3. GRAPHS WITH HIGH GIRTH AND CHROMATIC NUMBER

Let G be a graph. The clique number $\omega(G)$ and the independence number $\alpha(G)$ are defined in section 1. The *chromatic number* $\chi(G)$ is defined as the minimal number of colors necessary to color the vertices such that adjacent vertices have distinct colors. We call such a coloring *good*. In any good coloring the points of a clique are all distinct. Thus $\chi(G) \geq \omega(G)$. We might, *a priori*, think that equality should hold. In fact, the "opposite" may hold which can be seen in the following result. This result was proven independently by several authors [11, 18, 20].

THEOREM 3.1: *For every k there exists a finite graph G_k with*

$$\omega(G_k) = 2, \chi(G_k) \geq k.$$

158   Joel Spencer

Proof: The proof is by induction on k. For $k = 3$, G_k may be a pentagon. Assume the result for k and let $H_1, \ldots, H_k$ be k disjoint copies of G_k. For each ordered n-tuple $(\alpha_1, \ldots, \alpha_k)$, where $\alpha_i \in V(H_i)$, add a point $P = P_{\alpha_1 \ldots \alpha_k}$. This point P is adjacent only to the points α_i, $1 \le i \le k$. This graph is the desired G_{k+1}.

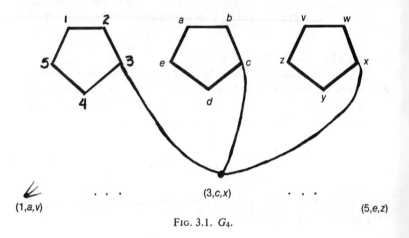

FIG. 3.1. G_4.

It is clear that $\omega(G_{k+1}) = 2$. A good k-coloring of G_{k+1} would give, since $\chi(G_k) \ge k$, all colors to each H_i. For $1 \le i \le k$ we find $\alpha_i \in H_i$ colored c_i. Then the point $P_{\alpha_1 \ldots \alpha_n}$ cannot be colored. Therefore, $\chi(G_{k+1}) \ge k + 1$.

Let $g(G)$, the *girth* of a graph G, denote the length of the minimal cycle of G. Theorem 3.1 gives a G of girth 4 and arbitrarily high chromatic number.

THEOREM 3.2 (Erdös [5]): *For all k, l there exists G*

$$\chi(G) \ge k, \quad g(G) \ge l.$$

This is a highly unintuitive result. If the girth is large, one can construe no reason why the graph could not be colored with a few colors. *Locally* it is easy to 3-color such a graph. However, we shall prove $\chi(G)$ is large due to *global* considerations. Basically, we note that if G is k-colored, some color is used at least n/k times ($n =$

$|V(G)|$). The vertices of that color form an independent set. Thus $\alpha(G) \geq n/\chi(G)$ or, in another form

$$\chi(G) \geq n/\alpha(G). \tag{3.1}$$

Now let $\mathbf{G} = \mathbf{G}_{n,p}$ be a random graph on vertex set $[n]$ with $P[\{i, j\} \in \mathbf{G}] = p$, with these probabilities mutually independent. If $S \in [n]^t$,

$$P[S \text{ is independent}] = (1 - p)^{\binom{t}{2}} \sim e^{-p\binom{t}{2}} \tag{3.2}$$

for p small.

$$P[\alpha(\mathbf{G}) \geq t] = P[\text{some } S \in [n]^t \text{ is independent}] \tag{3.3}$$

$$\leq \binom{n}{t} e^{-p\binom{t}{2}} \leq n^t e^{-pt^2/2}.$$

Setting $u = n/t$

$$P[\chi(\mathbf{G}) \leq u] \leq P[\alpha(\mathbf{G}) \geq t] \tag{3.4}$$

$$\leq [ne^{-pn/2u}]^{n/u}.$$

If p is somewhat larger than $(2u \ln n)/n$, then $\chi(\mathbf{G})$ is almost always greater than u. For u fixed (say $u = 100$), it is surprising how few edges $\left(\binom{n}{2} p \sim 100 n \ln n\right)$ we need before $\chi(\mathbf{G})$ is usually greater than u.

The obvious approach now would be to pick $p > (2k \ln n)/n$ so that $\chi(\mathbf{G}) > k$, and to show that $g(\mathbf{G})$ is large. However, this does not work because $\mathbf{G}$ has an expected number of triangles $\binom{n}{3} p^3 \sim$ (8/6) $k^3 (\ln n)^3$ and therefore almost always has girth 3. While $\mathbf{G}$ contains triangles and other small cycles, it has very few of them. An alteration is needed, three inches off the waist. We shall pick p such that $\mathbf{G}$ has few cycles, large chromatic number, and $\chi(\mathbf{G})$ remains large even after the removal of the cycles.

Proof of Theorem 3.2 (Erdös [5]): Fix k, l and let n be "sufficiently large" such that the approximations we make will be valid. Fix η, θ, $\theta < 1/l$, $0 < \eta < \theta/2$. Now look at **G** with $p = n^{\theta-1}$. Set $x = [n^{1-\eta}]$. We have seen that every $S \in [n]^x$ contains edges. Now we show that every $S \in [n]^x$ has at least n edges. We let $|\mathbf{G}|S|$ denote the number of edges of **G** restricted to S.

$$P[\text{some } S \in [n]^x \text{ has } |\mathbf{G}|S| \le n] \le \binom{n}{x}\binom{\binom{x}{2}}{n}(1-p)^{\binom{x}{2}}. \quad (3.5)$$

We bound

$$\binom{n}{x} \le n^n, \quad \binom{\binom{x}{2}}{n} \le n^{2n},$$

and approximate

$$(1-p)^{\binom{x}{2}} \sim e^{-px^2/2} \sim e^{-[n^{1+(\theta-2\eta)}]/2}.$$

Since $\theta - 2\eta > 0$, this "hyperexponential" factor dwarfs the n^n, n^{2n} terms. Thus

$$P[\text{every } S \in [n]^x \text{ has } |\mathbf{G}|S| > n] = 1 - o(1). \quad (3.6)$$

Now define $c_i(G)$ = the number of cycles in G of size i and $c(G) = \sum_{i=3}^{l} c_i(G)$. There are $(n)_i/2i$ possible i-cycles in **G**.* Each is in **G** with probability p^i. Therefore,

$$E[c_i(\mathbf{G})] = [(n)_i/2i]\, p^i < (np)^i \quad (3.7)$$

and thus

---

*We can associate ordered i-tuples $(x_1, \ldots, x_i)$ of distinct vertices with i-cycles $x_1 x_2 \ldots x_i x_1$. Since every cycle is associated with exactly $2i$ i-tuples the number of cycles is $(n)_i/2i$.

$$E[c(\mathbf{G})] = \sum_{i=3}^{l} E(c_i(\mathbf{G}))$$

$$\leq \sum_{i=3}^{l} (np)^i < l(np)^l. \tag{3.8}$$

From (3.8)

$$P[c(\mathbf{G}) > 2l(np)^l] < \tfrac{1}{2}. \tag{3.9}$$

By our choice of p, $2l(np)^l = 2ln^{\theta l} < n$. Combining (3.9), (3.6) we find a specific G satisfying

(a) $|G|S| \geq n$ for all $S \in [n]^x$,
(b) $c(G) < n$.

Let G^* be G with one edge deleted (arbitrarily) from each cycle of length $\leq l$. Since fewer than n edges are deleted

(a') $|G^*|S| > 0$ for all $S \in [n]^x$,
(b') G^* has no cycles of length $\leq l$.

Equation (3.1) and (a') yield

$$\chi(G^*) \geq n/x \geq n^\eta. \tag{3.10}$$

Because η depends only on l (it can be picked as any positive real less than $1/2l$), n^η can be arbitrarily large. We choose n sufficiently large such that $\chi(G^*) \geq k$.

Theorem 3.2 is one of the most aesthetically appealing uses of the probabilistic method. The statement of the result appears to call for purely constructive methods. Several years after the above proof was published, a quite difficult constructive proof was found by L. Lovász [12].

4. RANDOM GRAPHS

Here we examine the probable structure of the random graph $\mathbf{G} = \mathbf{G}_{n,p}$ which has vertex set $[n]$ and $P[\{i, j\} \in \mathbf{G}] = p$. These probabilities are mutually independent for all edges $\{i, j\}$. We will study the *evolution* of **G** as p increases. Random graphs have proven applicable to many diverse fields. Sociologists, modeling group interaction, identify a set of n individuals, each pair of individuals forming friendships with probability p. Assuming joint independence of the potential friendships, the "friendship graph" is $\mathbf{G} = \mathbf{G}_{n,p}$. Here p is a measure of gregariousness, and graph theoretic concepts such as connectivity and clique number have clear sociological import. This model might not be realistic—perhaps we should insert a "transitivity bias". For example, if i, j are friends and j, k are friends then i and k are more likely to be friends. Clearly, substantial theoretical work along these lines remains to be done.

Communication or transportation networks can also be represented by a model utilizing random graphs. If we wish to model an actual situation by a random graph, $\mathbf{G}_{n,p}$ should be replaced by a more complex model. Transportation networks may well require a negative transitivity bias. For example, if j has already been joined to i and k, there would be less need to join i to k. Finally, for an accurate model, the probabilities of the various edges might not be equal.

If A is a property that a graph may or may not possess, we denote by $P_{n,p}(A)$ the probability that **G** satisfies property A. We call a function $p(n)$ a *threshold function* for property A if

$$\lim P_{n, r(n)}(A) = \begin{cases} 0, \text{ if } \lim r(n)/p(n) = 0, & (4.1\text{a}) \\ 1, \text{ if } \lim r(n)/p(n) = 1, & (4.1\text{b}) \end{cases}$$

In this case we can perceive $f(p) = P_{n,p}(A)$ close to zero or nearly one for "almost all" p, "jumping" from zero to one when p is in the neighborhood of $p(n)$. Remarkably, many intrinsically interesting properties A (e.g., connectivity, nonplanarity) do possess a threshold function.

In this section we discuss the fundamental work of P. Erdös and A. Rényi [8]. They considered the evolution of $\mathbf{G}(n, e)$, a random graph on n vertices with e edges, as e ranges from 0 to $\binom{n}{2}$.

For all except the tightest results, the behavior of $\mathbf{G}(n, e)$ is nearly identical to $\mathbf{G}_{n,p}$, where $p = e/\binom{n}{2}$. We can approach the study of random graphs using either $\mathbf{G}(n, e)$ or $\mathbf{G}_{n,p}$. Both approaches have equal intuitive and practical appeal. We shall retain $\mathbf{G}_{n,p}$ because it facilitates the presentation of the proofs.

We will examine the behavior of $P_{n,p}(A)$ for various fundamental properties A. First, let A be the existence in G of a path of length k. Here k is a fixed constant. The result can be generalized to the appearance of other small subgraphs.

THEOREM 4.1: $p(n) = n^{-(k+1)/k}$ *is a threshold function for* A.

Proof: Enumerate all potential paths P_i, $1 \le i \le (n)_{k+1}/2$.* Set $M = (n)_{k+1}/2$ for convenience. For each i define an indicator random variable

$$\mathbf{X}_i = \begin{cases} 1, & \text{if } P_i \text{ is a path in } \mathbf{G}, \\ 0, & \text{if not,} \end{cases} \quad (4.2)$$

and set

$$\mathbf{X} = \sum_{i=1}^{M} \mathbf{X}_i, \quad (4.3)$$

the number of paths of length k in $\mathbf{G}$. Clearly,

$$E(\mathbf{X}_i) = P[P_i \text{ is a path in } \mathbf{G}] = p^k, \quad (4.4)$$

thus, by the linearity of expected values:

$$E(\mathbf{X}) = \Sigma E(\mathbf{X}_i) = Mp^k < n^{k+1}p^k. \quad (4.5)$$

---

*We can associate ordered $(i + 1)$-tuples $(x_1, \ldots, x_{i+1})$ of distinct vertices with paths $x_1 x_2 \cdots x_{i+1}$ of length i. Each path is associated with exactly two such $(i + 1)$-tuples $((x_1, \ldots, x_{i+1})$ and $(x_{i+1}, \ldots, x_1))$ so the number of such paths is $(n)_{i+1}/2$.

Now (4.1a) readily follows. For suppose $p = r(n)$ with

$$\lim_n r(n)n^{-(k+1)/k} = 0.$$

Then

$$\lim_n E(\mathbf{X}) = 0. \tag{4.6}$$

Since

$$P[\mathbf{G} \text{ satisfies } A] = P[\mathbf{X} > 0] \le E(\mathbf{X}), \tag{4.7}$$

we deduce $\lim_n P_{n,p}(\mathbf{G}) = 0$, as desired. In general, the proof of (4.1a) is easier than (4.1b).

Let us attempt to reverse the proof of (4.1a) to show (4.1b). Let $p = r(n)$ with

$$\lim_n r(n) \, n^{-(k+1)/k} = \infty. \tag{4.8}$$

Then

$$\lim_n E(\mathbf{X}) = \infty. \tag{4.9}$$

From (4.8) we cannot, *a priori*, deduce

$$\lim_n P(\mathbf{X} > 0) = 1. \qquad (?)(4.10)$$

The deduction of (4.10), from which (4,1b) readily follows, requires a more detailed look at the distribution of $\mathbf{X}$. We use the *second moment method* which requires a bound on the *variance* of $\mathbf{X}$.

We begin by squaring (4.3) and taking expectations.

$$\begin{aligned} E(\mathbf{X}^2) &= \sum_{i,j=1}^{M} E(\mathbf{X}_i \mathbf{X}_j) \\ &= M \sum_{j=1}^{M} E(\mathbf{X}_1 \mathbf{X}_j) \end{aligned} \tag{4.11}$$

by the obvious symmetry among the $\mathbf{X}_i$. For any j

$$E(\mathbf{X}_1 \mathbf{X}_j) = P[P_1 \text{ and } P_j \text{ are paths in } \mathbf{G}]$$
$$= p^{2k-\delta}, \qquad (4.12)$$

where δ is the number of common edges in the paths P_1, P_j.

Essentially, Var($\mathbf{X}$) will be small because most $\mathbf{X}_i$, $\mathbf{X}_j$ are independent, i.e., δ is usually zero. Let d_δ be the number of j for which P_j intersects P_1 in δ edges.

LEMMA: *For* $\delta > 0$, $d_\delta < \binom{k}{\delta}\binom{n}{k-\delta}((k+1)!/2))$.

Proof of Lemma: There are $\binom{k}{\delta}$ choices of the edge intersection of P_1 and P_j. The determination of δ edges in P_j defines *at least* $\delta + 1$ points of P_j. Now the point set of P_j (of size $k + 1$) can be completed in less than $\binom{n}{k-\delta}$ ways. Given the point set, there are at most $(k+1)!/2$ paths.

The bound on d_δ is perhaps a gross overestimate. However, all we really need is

$$d_\delta < c_k n^{k-\delta}, \qquad (4.13)$$

where c_k depends only on k. Now

$$\sum_{j=1}^{M} E(\mathbf{X}_1 \mathbf{X}_j) = \sum_{\delta=0}^{k} d_\delta p^{2k-\delta}$$
$$< Mp^{2k} + \sum_{\delta=1}^{K} c_k n^{k-\delta} p^{2k-\delta} \qquad (4.14)$$

because $d_0 \leq M$. From (4.11) (using the approximation $M \sim n^{k+1}/2$)

$$E(\mathbf{X}^2) = M \sum_{j=1}^{M} E(\mathbf{X}_1 \mathbf{X}_j)$$
$$< M^2 p^{2k} \left[1 + \sum_{\delta=1}^{k} 2c_k n^{-1}(pn)^{-\delta} \right]. \qquad (4.15)$$

Equations (4.8), (4.15) yield

$$\lim_n \sum_{\delta=1}^{k} 2c_k n^{-1}(pn)^{-1} = 0. \tag{4.16}$$

Therefore, since $E(\mathbf{X})^2 = M^2 p^{2k}$,

$$\lim_n E(\mathbf{X}^2)/E(\mathbf{X})^2 = 1. \tag{4.17}$$

The remainder of the proof is pure probability theory using (4.9) and (4.17) to imply (4.10). Since $\text{Var}(\mathbf{X}) = E(\mathbf{X}^2) - E(\mathbf{X})^2$,

$$\lim_n \text{Var}(\mathbf{X})/E(\mathbf{X})^2 = 0. \tag{4.18}$$

Chebyshev's inequality states that for any $\mathbf{X}$ with mean μ, variance σ^2 and any $\lambda > 0$

$$P[|\mathbf{X} - \mu| \geq \lambda \sigma] \leq \lambda^{-2}. \tag{4.19}$$

If $\mathbf{X}$ is non-negative integral valued, this implies

$$P[\mathbf{X} = 0] \leq [\mu/\sigma]^{-2} = \text{Var}(\mathbf{X})/E(\mathbf{X})^2. \tag{4.20}$$

Thus (4.18) implies (4.10), concluding the proof of Theorem 4.1.

While the second moment method generally has wide application, there exist examples where this is not the case. Call G Hamiltonian if there is a cycle passing through every point exactly once. Let property A be "$\mathbf{G}$ is Hamiltonian". Setting $\mathbf{X}$ equal to the number of Hamiltonian circuits in G we find $E(\mathbf{X}) = (n!/2)p^n$. If $p = c/n$ where $c > e$, then $\lim E(\mathbf{X}) = \infty$. However, $p(n) = n^{-1}$ is not a threshold function for $\mathbf{G}$. For if $p = r(n) = o(n^{-1} \ln n)$, then $\mathbf{G}$ will have isolated points with probability $1 - o(1)$ and therefore not be Hamiltonian. It has recently been shown by the Hungarian mathematician Posa [15] that $p(n) = n^{-1} \ln n$ is a threshold function for being Hamiltonian.

The evolution of $\mathbf{G}$ paints an interesting and instructive picture. Set $p = n^\alpha$. As α approaches -1 from below, the maximal path length of **G** increases. When α reaches -1 and $p = cn^{-1}$, the components of **G** begin to congeal into larger groups. The critical point is at $c = 1$, $p = n^{-1}$. For $c < 1$ the largest component has size $k(c) \ln n$ but for $c > 1$ the largest component size jumps to $k'(c)n$. This jump can be explained by considering the component of **G** containing a given point as the result of a "branching process". Each point branches, on the average, to $c = pn$ others. If $c < 1$, the process quickly dies. But if $c > 1$, the process may grow to absorb much of the graph.

Planarity also changes at $c = 1$. Let us assume $c < 1$ and show that **G** is planar. The expected number of cycles of size t is

$$[(n)_t/2] p^t < (np)^t = c^t.$$

Consequently the expected number of points lying on cycles is $< \sum_{t=3}^{n} tc^t = k$, a constant depending only on c. Set $w = w(n) = \ln n$. (Actually, $w(n)$ can be any function slowly approaching infinity.) Then

$$P[\geq w \text{ points of } \mathbf{G} \text{ are in cycles}] = o(1). \quad (4.21)$$

Also

$$P[\text{some } w \text{ points of } \mathbf{G} \text{ have } > w \text{ edges}] \leq \binom{n}{w}\binom{\binom{w}{2}}{w+1} p^{w+1}$$

$$\leq n^w w^{2w} p^{w+1} = (cw^2)^w p = o(1). \quad (4.22)$$

That is, at most w points of **G** are in cycles and no w points have more than w edges. Thus **G** cannot have *intersecting* cycles and, by elementary graph theory, is planar.

We can further demonstrate that if $c > 1$, **G** is not planar.

As c increases beyond one, the components of **G** merge. For large c nearly all the points lie in a single component. However,

$$n(1 - p)^{n-1} \sim e^{-c}n$$

points remain isolated. Full connectivity is achieved at $p(n) = n^{-1} \ln n$. There is a surprisingly precise result.

THEOREM 4.2: $n^{-1} \ln n$ *is a threshold function for connectivity. Moreover, setting* $p = p(n) = n^{-1} \ln n + a\, n^{-1}$,

$$\lim_n P[\mathbf{G}_{n,p} \text{ is connected}] = e^{-e^{-a}}. \qquad (4.23)$$

We are content to indicate the proof of (4.23). **G** may have isolated points but it will not have small components. The probability that **G** has a two point component is bounded by $\binom{n}{2}(1-p)^{2(n-2)}\,p$, the factors representing the number of potential 2 point components and the probabilities that the two points will be nonadjacent to all other points and adjacent to each other. The factor p makes this expression $o(1)$. In general, G will have no k point components for $k \leq n/2$ (actually $k < n(1-\epsilon)$).

Now we can concentrate on isolated points. Any particular point is isolated with probability $(1-p)^{n-1} \sim e^{-a}$. Let **Y** be the number of isolated points. We can show that **Y** is, in the limit, Poisson (we suppress the elementary, albeit nontrivial, details) and thus

$$\begin{aligned} P[\mathbf{G} \text{ connected}] &= o(1) + P[\mathbf{Y} = 0] \\ &= o(1) + e^{-e^{-a}}. \end{aligned} \qquad (4.24)$$

5. CODING THEORY

The most celebrated use of the probabilistic method occurs not in combinatorial mathematics but rather in information theory. In this section we outline the result of Claude Shannon which contends that it is possible to transmit information through a "noisy channel" at a "positive rate" with the probability of correct reception near unity.

Coding theory boasts a voluminous literature. We mention only a few general references: Berklekamp [1], Peterson and Weldon [14], and van Lint [19].

Electrical engineers are primarily responsible for the development of coding theory and they are wont to make diagrams like Figure 5.1.

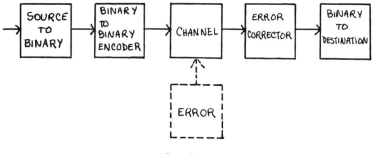

Fig. 5.1.

Our general problem is to transmit information through a channel.* This information is originally in some language, e.g., English, utilizing a standard set of alphanumeric characters, and we must first translate it to binary *bits* (i.e., 0's and 1's) by some method, e.g., Morse Code. The final step will be to translate the binary message back to the original language. We won't concern ourselves with the initial or the terminal steps here. Instead, we assume the message is originally in binary, leaving the more restricted network of Figure 5.2.

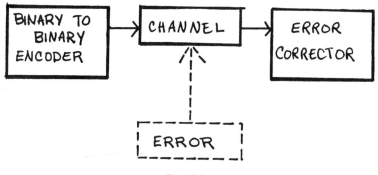

Fig. 5.2.

---

*We usually think of the channel as a telephone or telegraph line.

Information is sent into the channel in bits. We make the basic assumption* that in the channel each bit is transmitted correctly with probability $1 - p$ and incorrectly (i.e., "switched") with probability p and that these probabilities are mutually independent. Here p, called the error probability, is a parameter of the channel. Note that if $p = 0$, no encoding or decoding is necessary—the channel is perfect. If $p = \frac{1}{2}$, it is impossible to transmit information since, regardless of what is transmitted, a uniformly distributed message is received. We assume $0 < p < \frac{1}{2}$. If $p > \frac{1}{2}$, the receiver may first switch every bit, effectively changing p to $1 - p$.

Let us fix p, say $p = .1$, and consider various schemes for transmitting information. We could send each bit three times (i.e., the binary encoder maps x into xxx) and the error corrector would translate a block of three bits by the bit appearing most often. Letting p_E denote the probability of incorrect reception, a simple calculation shows

$$p_E = p^3 + 3p^2(1 - p) = .028$$

which is considerably better than .1. We have lost "speed," for a message of length n requires $3n$ bits to be sent. Define the *rate of transmission* $= r_T$ as the number of bits sent divided by the number of bits in the original message. In this case, $r_T = 1/3$.

If we sent each bit 9 times, then $r_T = 1/9$ and $p_E \sim .0003$. Clearly, we can make p_E arbitrarily small, but at the sacrifice of making r_T small. Shannon proved that this tradeoff was not necessary! Define the *entropy* function

$$H(p) = -p \log_2 p - (1 - p) \log_2(1 - p). \tag{5.1}$$

Shannon's Theorem states that there exist codes with p_E arbitrarily small and r_T arbitrarily close to $1 - H(p)$. Since $H(.1) \sim .4610$,

---

*Many other possible assumptions (e.g., that errors are not independent but are likely to occur in "bursts," or that 1's become 0's but 0's never become 1's) have been studied in great detail.

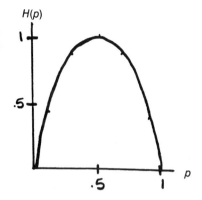

FIG. 5.3. Entropy curve and sample values.

for $p = .1$, we can send information with arbitrary accuracy at a rate of nearly 54%. For $p = .49$, $H(p) \sim .999711$, so the rate of transmission is only .0289%. Shannon also proved that if arbitrary accuracy is required, we cannot do better than rate $1 - H(p)$, but we do not show that here.

It will be necessary to place a structure on the set of possible messages. Let C_n denote the n-dimensional vector space over $Z/2$ with a fixed basis. The 2^n points of C_n are of the form $(\epsilon_1, \ldots, \epsilon_n)$, $\epsilon_i = 0$ or 1, and represent possible messages of length n. Let $\vec{x} = (x_1, \ldots, x_n)$, $\vec{y} = (y_1, \ldots, y_n)$ be vectors in C_n. We define a distance function, called the *Hamming Metric*, on C_n by

$$\rho(\vec{x}, \vec{y}) = \sum_{i=1}^{n} |x_i - y_i| \quad \text{(NOT modulo 2)}. \tag{5.2}$$

This is, $\rho(\vec{x}, \vec{y})$ is the number of coordinates by which $\vec{x}$ and $\vec{y}$ differ. The reader may easily check that ρ does indeed satisfy the triangle inequality

$$\rho(\vec{x}, \vec{z}) \leq \rho(\vec{x}, \vec{y}) + \rho(\vec{y}, \vec{z}).$$

We also define a weight function

$$w(\vec{x}) = \sum_{i=1}^{n} x_i \quad \text{(NOT modulo 2)}. \tag{5.3}$$

Note

$$w(\vec{x}) = \rho(\vec{x}, \vec{0}),$$

$$w(\vec{x} - \vec{y}) = \rho(\vec{x}, \vec{y}). \quad (5.4)$$

Let $S \subseteq C_n$, $|S| = 2^k$. We describe a method of transmitting information which we call the *S-code*. Fix a bijection $\psi: C_k \to S$. Our encoder will split the original message (assumed arbitrarily long) into blocks of length k. For each block, which is a vector $\vec{x} \in C_k$, the codeword $\vec{y} = \psi(x) \in S \subseteq C_n$ is sent into the channel. This gives a rate of transmission k/n. For example, sending each bit three times corresponds to $k = 1$, $n = 3$, $S = \{(0, 0, 0), (1, 1, 1)\}$. In the channel the message $\vec{y}$ is changed to $\vec{y} + \vec{e}$, vector addition modulo two. The vector $\vec{e} = (\epsilon_1, \ldots, \epsilon_n)$ is called the *error vector*. Our basic assumption on the nature of the channel is that $P[\epsilon_i = 1] = p$, $1 \le i \le n$, and these probabilities are mutually independent.

The decoder receives the vector $\vec{y} + \vec{e} \in C_n$. It finds the $\vec{s} \in S$ which minimizes $\rho(\vec{s}, \vec{y} + \vec{e})$. If more than one $\vec{s}$ assumes the minimum, then the decoder uses an arbitrary device to select one. The decoder assumes $\vec{s}$ has been transmitted (we will say the decoder *corrects to* $\vec{s}$) and decodes as $\psi^{-1}(s)$. Correct decoding occurs iff $\vec{s} = \vec{y}$.

The S-code has rate of transmission $r(S) = \log_2(|S|)/n$. For arbitrary $|S|$, say $2^k \le |S| < 2^{k+1}$, we can modify the S-code by letting $\psi: C_k \to S$ be injective. Then

$$r(S) = [\log_2(|S|)]/n. \quad (5.5)$$

We define $p_E(S)$ as the probability of error in the S-code. Of course, p_E depends on p.

Before giving Shannon's proof, which requires nonconstructive methods, we consider one more example of a constructive code. Let us assume $p = 10^{-6}$ and set $N = 31$. Consider the 5×31 matrix

$$A = \begin{bmatrix} 0 & 0 & \cdot & \cdot & \cdot & 1 & 1 \\ 0 & 0 & \cdot & \cdot & \cdot & 1 & 1 \\ 0 & 0 & \cdot & \cdot & \cdot & 1 & 1 \\ 0 & 0 & \cdot & \cdot & \cdot & 1 & 1 \\ 1 & 0 & \cdot & \cdot & \cdot & 0 & 1 \end{bmatrix} \quad (5.6)$$

whose columns are all possible 0-1 vectors, not all zero. Let $\vec{c}_i$, $1 \le i \le 31$, denote the ith column vector. Define

$$S = \{\vec{x} \in C_{31}: A\vec{x} = \vec{0}\}, \quad (5.7)$$

where the matrix multiplication is in $Z/2$. Since A has rank 5, $|S| = 2^{31-5}$ and hence $r(S) = 26/31$.

We claim that S has minimal weight 3. That is, if $\vec{s} \in S$, $\vec{s} \ne \vec{0}$, then $w(\vec{s}) \ge 3$. For suppose $\vec{s} = (\epsilon_1, \ldots, \epsilon_n)$ where $\epsilon_i = \epsilon_j = 1$, $\epsilon_k = 0$ for $k \ne i, j$. Then $A\vec{s} = \vec{0}$ implies $\vec{c}_i + \vec{c}_j = \vec{0}$ so $\vec{c}_i = \vec{c}_j$, which is impossible. It is even simpler to show that $w(\vec{s})$ cannot be unity. Now we take advantage of the fact that S is a subspace. Let $\vec{s}, \vec{t} \in S$, $\vec{s} \ne \vec{t}$. Then $\vec{s} - \vec{t} \in S$ so

$$\rho(\vec{s}, \vec{t}) = w(\vec{s} - \vec{t}) \ge 3.$$

This means that the S-code is 1-error correcting. That is, if $\vec{s}$ is sent and $\vec{s} + \vec{e}$ received where $w(\vec{e}) \le 1$, then decoding will be correct. For if the error corrector corrects $\vec{s} + \vec{e}$ as $\vec{t} \in S$, then

$$\rho(\vec{t}, \vec{s} + \vec{e}) \le \rho(\vec{s}, \vec{s} + \vec{e}) \le 1,$$

so

$$\rho(\vec{t}, \vec{s}) \le \rho(\vec{t}, \vec{s} + \vec{e}) + \rho(\vec{s} + \vec{e}, \vec{s}) \le 2$$

and therefore $\vec{t} = \vec{s}$.

We now bound $p_E(S)$ by the probability of more than one error in 31 bits, *viz.*,

$$\rho_E(S) \leq \binom{31}{2} p^2 = 4.65 \times 10^{-10},$$

a considerable improvement on 10^{-6}. This is an example of a *Hamming Code*, which is widely applied to practical problems.

Now we outline the proof of Shannon's Theorem. Fix the error rate p and an arbitrary $\epsilon > 0$. Let n be very large. Let

$$\mathbf{S} = \{\vec{s}_0, \vec{s}_1, \ldots, \vec{s}_M\},$$

where $M = 2^{n(1-H(p)-\epsilon/2)}$ and the $\vec{s}_i$, $0 \leq i \leq M$, are chosen independently with uniform distribution on C_n. We may have $\vec{s}_i = \vec{s}_j$ for some i, j. With probability $1 - o(1)$, $|\mathbf{S}| = M(1 - o(1))$ and therefore

$$r(S) = 1 - H(p) - \epsilon/2 - o(1) \geq 1 - H(p) - \epsilon. \tag{5.8}$$

We show $p_E(\mathbf{S})$ is small by examining incorrect decoding occurrences in the S-code.

By symmetry we can assume $\vec{s}_0$ is transmitted. The message $\vec{s}_0 + \vec{e}$ is received. Incorrect translation occurs if for some i, $1 \leq i \leq M$,

$$\rho(\vec{s}_i, \vec{s}_0 + \vec{e}) \leq \rho(\vec{s}_0, \vec{s}_0 + \vec{e}) = w(\vec{e}). \tag{5.9}$$

The weight $w(\vec{e})$ has binomial distribution with parameters n, p. By the weak law of large numbers

$$P[w(\vec{e}) \geq n(p + \delta)] = o(1), \tag{5.10}$$

for any fixed $\delta > 0$. We choose δ such that

$$H(p + \delta) \leq H(p) + \epsilon/4. \tag{5.11}$$

If incorrect decoding occurs, then either

(a) $w(\vec{e}) \geq n(p + \delta)$

or

(b) $\rho(\vec{s}_i, \vec{s}_0 + \vec{e}) \le n(p + \delta)$ for some i, $1 \le i \le M$.

The probability of (a) occurring is $o(1)$. Let

$$\mathbf{B} = B(\vec{s}_0 + \vec{e}, n(p + \delta)) = \{\vec{x} \in C_n : \rho(\vec{x}, \vec{s}_0 + \vec{e})\ n(p + \delta)\} \tag{5.12}$$

denote the ball of radius $n(p + \delta)$ with center $\vec{s}_0 + \vec{e}$. Then (b) is equivalent to

(b') $\vec{s}_i \in \mathbf{B}$ for some i, $1 \le i \le M$.

Now

$$|\mathbf{B}| = \sum_{i=0}^{n(p+\delta)} \binom{n}{i} \le n\binom{n}{n(p + \delta)}. \tag{5.13}$$

By an elementary application of Stirling's Formula

$$\binom{n}{n\alpha} = 2^{n(H(\alpha)+o(1))} \text{ for } 0 < \alpha < 1. \tag{5.14}$$

(This is the origin of the function $H(p)$.) Hence

$$|\mathbf{B}| = n\ 2^{n(H(p)+\delta+o(1))} \le 2^{n(H(p)+\epsilon/4+o(1))}. \tag{5.15}$$

For $1 \le i \le M$ the point $\vec{s}_i$ is *jointly independent* of $\vec{s}_0$ and $\vec{e}$, and therefore *independent* of $\mathbf{B}$. Thus

$$P[\vec{s}_i \in \mathbf{B}] = |\mathbf{B}|/|C_n| = 2^{n(H(p)+\epsilon/4+o(1)-1)} \tag{5.16}$$

and

$$P[(b')] = P[\vec{s}_i \in \mathbf{B} \text{ for some } 1 \le i \le M]$$

$$\le M\ 2^{n(H(p)+\epsilon/4+o(1)-1)}$$

$$= 2^{n(-\epsilon/4+o(1))} = o(1). \tag{5.17}$$

Since the probabilities of (a) and (b′) are both small, correct decoding occurs with probability near unity. Consequently, the random S-code has $r_T \geq 1 - H(p) - \epsilon$ and $p_E = o(1)$ almost always, and there must exist *specific* good S-codes—proving Shannon's Theorem.

The nonconstructive nature of the proof of Shannon's Theorem leads to important practical problems. To understand these, we first consider *implementation* of encoders and decoders for the Hamming Code. The S defined by (5.7) is the set of $(\epsilon_1, \ldots, \epsilon_{31}) \in C_{31}$ satisfying five independent linear equations. When the appropriate variables are isolated these equations become

$$\epsilon_1 = \epsilon_3 + \epsilon_5 + \epsilon_7 + \cdots + \epsilon_{31},$$

$$\epsilon_2 = \epsilon_3 + \epsilon_6 + \epsilon_7 + \epsilon_{10} + \epsilon_{11} + \cdots + \epsilon_{30} + \epsilon_{31},$$

$$\epsilon_4 = \epsilon_5 + \epsilon_6 + \epsilon_7 + \epsilon_{12} + \epsilon_{13} + \epsilon_{14} + \epsilon_{15} + \cdots + \epsilon_{28} + \epsilon_{29} + \epsilon_{30} + \epsilon_{31}$$

$$\epsilon_8 = \epsilon_9 + \cdots + \epsilon_{15} + \epsilon_{24} + \cdots + \epsilon_{31}$$

$$\epsilon_{16} = \epsilon_{17} + \cdots + \epsilon_{31}. \tag{5.18}$$

Now we may encode by a map $\psi: C_{26} \to C_{31}$. We define

$$\psi((\delta_1, \ldots, \delta_{26})) = (\epsilon_1, \ldots, \epsilon_{31})$$

by $\epsilon_3 = \delta_1$, $\epsilon_5 = \delta_2$, $\epsilon_6 = \delta_3$, $\epsilon_7 = \delta_4$, (skipping the ϵ_{2i}) and $\epsilon_1, \epsilon_2, \epsilon_4, \epsilon_8, \epsilon_{16}$, by (5.18). Implementing this linear encoding with electronic circuitry presents no problems.

Similarly, decoding is easy to apply. Assume that a vector $\vec{u}$ is received by the error corrector. Calculate $A\vec{u} = \vec{w} \in C_5$. If $\vec{w} = \vec{0}$, then $\vec{u} \in S$, so we correct to $\vec{u}$. If $\vec{w} \neq 0$, we need find $\vec{e_i}$ (all coordinates zero except the ith coordinate one) such that $A(\vec{u} + \vec{e_i}) = 0$. But

$$A(\vec{u} + \vec{e_i}) = \vec{0} \text{ iff } \vec{w} = A\vec{u} = A\vec{e_i} = \vec{c_i}.$$

NONCONSTRUCTIVE METHODS IN DISCRETE MATHEMATICS 177

Since $\vec{c}_i$, $1 \leq i \leq 31$, are all possible nonzero vectors of C_5, we solve for i. In fact, if $u = (u_1, \ldots, u_5)$, then i has binary expansion $u_1 u_2 u_3 u_4 u_5$. We correct to $\vec{u} + \vec{e}_i$. Translation, the calculation of ψ^{-1}, merely requires the deletion of the 1st, 2nd, 4th, 8th, and 16th coordinates.

Suppose S is achieved from the nonconstructive proof of Shannon's Theorem. Even if a specific S is found it is likely to have no useful structure. This will make encoding and decoding dependent on lengthy tables. This is often impossible due to physical constraints. For one thing, the time required for table lookup might effectively decrease the rate by a substantial amount.

If S is a subspace of C_n, we call the S-code a *group code*. Group codes are considerably easier to encode and decode. Fortunately, we can adjust the proof of Shannon's Theorem to show that "good" group codes exist.

Let $\vec{v}_1, \ldots, \vec{v}_t$, $t = n(1 - H(p) - \epsilon)$, be independently uniformly selected from C_n. Let **S** be the subspace generated by the $\vec{v}_i$. We can show that the $\vec{v}_i$ are independent with probability $1 - o(1)$, so $r(\mathbf{S}) = 1 - H(p) - \epsilon$.

Assume code word $\vec{c} \in \mathbf{S}$ is transmitted and the error vector is $\vec{e}$. Incorrect translation occurs iff for some $\vec{d} \in S$, $\vec{d} \neq \vec{c}$, $\rho(\vec{d}, \vec{c} + \vec{e}) \leq \rho(\vec{c}, \vec{c} + \vec{e})$. This is equivalent to $\rho(\vec{d} - \vec{c}, \vec{e}) \leq \rho(\vec{0}, \vec{e}) = w(\vec{e})$. Here $\vec{d} - \vec{c}$ ranges over all nonzero vectors in S. Hence, the probability of incorrect translation is the same whether $\vec{c}$ or $\vec{0}$ is transmitted. Consequently, we may assume $\vec{0}$ is transmitted. The probability of incorrect decoding is bounded by $o(1)$ terms plus the probability that some $\vec{h} = h_1 \vec{v}_1 + \cdots + h_t \vec{v}_t$, not all h's zero, has weight less than $n(p + \delta)$. For any fixed $h_1, \ldots, h_t$, since the $\vec{v}_i$ are independent and uniformly distributed, $\vec{h}$ is *uniformly distributed* on C_n. Thus

$$P[w(\vec{h}) \leq n(p + \delta)] = |B(\vec{0}, n(p + \delta))| 2^{-n}$$
$$= 2^{n(h(p+\delta) - 1 + o(1))}, \quad (5.19)$$

and so

$$P[\text{some } \vec{h} \in \mathbf{S} - \{0\} \text{ has } w(\vec{h}) \leq n(p + \delta)]$$
$$\leq |\mathbf{S}| \, 2^{n(H(p+\delta) - 1 + o(1))} = o(1). \quad (5.20)$$

Completion of this proof is identical to the first proof.

The proof of the existence of good group codes, while of considerable interest, remains nonconstructive in nature. An important open problem in coding theory is to construct, for a given p, a sequence of codes with rates approaching $1 - H(p)$ and probability of correct reception approaching unity.

REFERENCES

1. E. R. Berklekamp, *Algebraic Coding Theory*, McGraw-Hill, New York, 1968.
2. H. Chernoff, "A measure of asymptotic efficiency for tests of a hypothesis based on the sum of observations," *Ann. Math. Statist.*, 23 (1952), 493-509.
3. M. Dwass, *Probability*, Benjamin, New York, 1970.
4. P. Erdös, "Some remarks on the theory of graphs, *Bull. Amer. Math. Soc.*, 53 (1947), 292-294.
5. ———, "Graph theory and probability", *Canad. J. Math.*, 11, 34-38.
6. P. Erdös and L. Lovász, "Problems and results on 3-chromatic hypergraphs and some related questions", *Finite and Infinite Sets*, North Holland, Amsterdam, 1975.
7. P. Erdös and J. W. Moon, "On sets of consistent arcs in a tournament," *Canad. Math. Bull.*, 8 (1965), 269-271.
8. P. Erdös and A. Renyi, "On the evolution of random graphs," *Mat. Kutato Int. Kozl.*, 5 (1960), 17-60.
9. P. Erdös and J. Spencer, *Probabilistic Methods in Combinatorics*, Academic Press, New York, 1974 (Akadémiai Kiadó).
10. R. L. Graham and J. Spencer, "A constructive solution to a tournament problem," *Canad. Math. Bull.*, (1) 14 (1971), 45-48.
11. J. B. Kelly and L. B. Kelly, "Paths and circuits in critical graphs", *Amer. J. Math.*, 76 (1954), 786-792.
12. L. Lovász, "On the chromatic number of finite set-systems", *Acta. Math. Acad. Sci. Hungar.*, (1-2) 19 (1968), 59-67.
13. J. W. Moon, *Topics on Tournaments*, Holt, Rinehart, and Winston, New York, 1968.
14. W. W. Peterson and E. J. Weldon, *Error-Correcting Codes*, M. I. T. Press, Cambridge, 1972.
15. L. Posa (private correspondence).
16. T. Szele, "Kombinatorikai vizsgalatok az iranyitott teljes graffal kaposolatban, *Mat. Fiz. Lapik*, 50 (1943), 223-256 (in Hungarian).
17. J. Spencer, Optimal Ranking of Tournaments, *Networks*, 1 (1971), 135-138.
18. W. Tutte, (under pseudonym Blanche Descartes), "A three-colour problem," *Eureka* (April 1947, solution March 1948) and "Solution to Advanced Problem No. 4526," *Amer. Math. Monthly*, 61 (1954), 352.
19. J. H. van Lint, *Coding Theory*, Springer-Verlag, Berlin, 1971.
20. A. A. Zykov, "On some properties of linear complexes" (in Russian), *Russian Math. Sbornik, N.S.*, 24 (66) (1949), 163-188.

MATROIDS AND COMBINATORIAL GEOMETRIES

Tom Brylawski* and Douglas G. Kelly

INTRODUCTION

The theory of combinatorial geometries, or matroids, began with Hassler Whitney's paper [45], in which he pointed out that many of the properties of linear dependence in vector spaces do not depend on the definition of dependence, but only on certain axioms, the so-called "abstract properties of linear dependence". Moreover, Whitney and others pointed out that these same axioms are satisfied by dependence relations arising in other areas, such as graph theory, transversal theory, and algebraic field theory. Because of its axiomatic approach to dependence relations, the theory has been called "coordinate-free linear algebra".

In this survey we present the subject informally by giving a few equivalent axiom systems, mentioning some of the primary examples, and indicating several of the current lines of research in the area. The reader desiring a more complete and rigorous treatment is referred to [45], [14], [41], [28], [52], and [11]. The last-named paper served as a basis for this survey; it includes numerous exer-

---

*Author was partially supported by N.S.F. Grant GP 42375.

cises and also contains a more complete bibliography than does this paper.

I. DEFINITIONS AND EXAMPLES

Independent sets; affine and dependence matroids. A *matroid* (*combinatorial pregeometry*) is a pair $M = (S, \mathcal{I})$, where S is a finite set of points and $\mathcal{I}$ is a nonempty family of subsets of S satisfying:

I1. If $A \subseteq B$ and $B \in \mathcal{I}$, then $A \in \mathcal{I}$,

I2. If $A, B \in \mathcal{I}$ and $|A| < |B|$, then there exists $b \in B - A$ with $A \cup \{b\} \in \mathcal{I}$.

Members of $\mathcal{I}$ are called *independent sets* of M; other subsets are called *dependent*. Two matroids are *isomorphic* if a bijection between their sets preserves independent and dependent subsets.

The simplest families $\mathcal{I}$ satisfying these axioms are the families $\mathcal{I}_k$, where $\mathcal{I}_k$ contains all sets of k or fewer points. For reasons that will emerge later, the matroid $(S, \mathcal{I}_k)$ is called the *free matroid* of rank k on S, and is denoted $F^{n,k}$, where n is the number of members of S. Special cases are $F^{n,0}$, the trivial matroid on n points, in which no nonempty set is independent, and $F^{n,n}$, the *free geometry* or *Boolean algebra*, in which every set is independent.

The prototypical examples of matroids are finite sets of vectors in vector spaces; if S is such a set, let $\mathcal{I}$ be the family of all linearly independent subsets of S. It is an exercise in elementary linear algebra to check that I1 and I2 are satisfied. These examples prompted Whitney's coining of the term "matroid", since any $m \times n$ matrix D over a field K can be viewed as a set of n (possibly nondistinct) vectors in m-dimensional vector space over K, viz., the columns of D. (Similarly, of course, the rows of a matrix are the points of a matroid.)

Example 1. Consider the matrix

$$D = \begin{bmatrix} 0 & 0 & 2 & 0 & 1 \\ 0 & 1 & 3 & 3 & 3 \\ 0 & 0 & 2 & 0 & 1 \end{bmatrix}$$

MATROIDS AND COMBINATORIAL GEOMETRIES 181

over the real numbers; if the columns of D are labeled a, b, c, d, e, then the independent sets of the matroid are $\emptyset$, b, c, d, e, bc, be, cd, ce, de. (We will sometimes use the convention of denoting finite sets without braces or commas; thus $abfg = \{a, b, f, g\}$.)

The matroid M arising from a matrix D in this way is called the (column) *dependence* matroid of D and D is said to *coordinatize* M over K. It is apparent that the geometric structure of $M(D)$ depends only on the row space of D and that any spanning subset of this row space (e.g., the first two rows of D in our example) would serve equally well to coordinatize M. The matroid above is also coordinatized by the matrix

$$D' = \begin{bmatrix} 0 & 0 & 1 & 0 & 1 \\ 0 & 1 & 0 & 1 & 1 \end{bmatrix}.$$

Another matroid on a finite set of points in a vector space is the *affine matroid*, in which $A \in \mathcal{I}$ if A is *affinely independent*, i.e., contains no 3 points on a line, no 4 points on a plane, and in general no $n + 2$ points on an n-dimensional flat (affine subspace). (Formally, $A = \{s_1, \ldots, s_r\}$ is affinely independent if $\sum_{i=1}^{r} a_i = 0$ and $\sum_{i=1}^{r} a_i s_i = 0$ imply that $a_i = 0$ for $i = 1, \ldots, r$. Axioms I1 and I2 follow easily from this formulation.) Affinely independent points are sometimes said to be in *general position* relative to each other.

Affine matroids have natural pictorial representations, as the following example shows.

Example 2. Let $S = \{a, b, c, d, e, f, g\}$ be a set of points in Euclidean 3-space arranged as shown in Figure 1.

The independent sets are the following:

the empty set and all singletons and pairs,
all 3-sets except cdg,
all 4-sets except $abef$ and sets containing cdg.

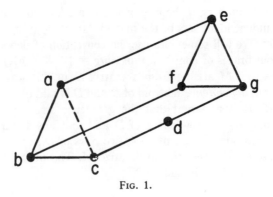

Fig. 1.

An affine matroid always has the property that singletons and pairs are independent; such a matroid is called a *combinatorial geometry*. We will shorten this term to *geometry*.

By the use of certain pictorial conventions, we can use "affine pictures" to represent many matroids which are not geometries; for example, the matroid of Example 1 can be pictured as

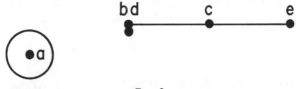

Fig. 2.

The conventions are these: if an element is enclosed in a circle, it is viewed as spanning no (affine) flat; if two or more elements are juxtaposed at a point, then any of them spans the 0-dimensional flat consisting of that point. The reader can check that $F^{n,k}$ is always a (real) affine geometry for $k \geq 2$, being pictured by n points in free position in $(k - 1)$-dimensional affine (Euclidean) space. The classical example of an affine geometry is the set of all q^n n-tuples of scalars from a finite field F_q. Such a geometry is denoted AG(n, q).

In dependence matroids and affine geometries, the notions of "flat", "dimension" or "rank", and "closure" suggest themselves.

For an arbitrary matroid M on a set S, we say that $a \in S$ *depends* on $A \subseteq S$ if either $a \in A$ or $B \cup a$ is dependent for some independent subset B of A. We define the *closure* $\overline{A}$ of a subset A of S as the set of all points that depend on A. The *closed sets*, or *flats*, of M are the sets A for which $\overline{A} = A$. We say that a flat F *covers* a flat G if $F \supset G$ and $F \supseteq H \supseteq G$ implies $H = F$ or $H = G$.

The characteristic properties of the family $\mathcal{F}$ of flats of a matroid are:

F1. $S \in \mathcal{F}$.

F2. If $A, B \in \mathcal{F}$, then $A \cap B \in \mathcal{F}$.

F3. For any $A \in \mathcal{F}$, $S - A$ is partitioned by the members of $\mathcal{F}$ that cover A (in that every $s \notin A$ is in one and only one flat covering A).

And it can be shown that if $\mathcal{F}$ is any family of subsets of S satisfying these axioms, then $\mathcal{F}$ is the family of flats of the matroid $(S, \mathcal{I})$, in which $A \in \mathcal{I}$ if and only if every proper subset of A is contained in some flat not containing A.

Examples. In Example 1 above, the flats are the sets a, abd, ac, ae, $abcde$. The geometry of Example 2 has the following sets as flats:

$\emptyset$ and all singletons,

all pairs except cd, cg, and dg,

cdg, and all other triples not contained in one of the three "rectangular faces of the prism",

abef, *acdeg*, *bcdfg*, and *abcdefg*.

Notice that a matroid is a geometry if and only if $\emptyset$ and all singletons are flats. If $\emptyset$ is not closed, then $\overline{\emptyset}$ contains the dependent singletons, which are called *loops* (the reason for this name will emerge below); these points are in every flat.

We say that A *spans* B if $\overline{A} = B$; the minimal spanning subsets of a flat B and the maximal independent subsets of B coincide and are called the *bases* (singular: *basis*) of B. The bases of S are called bases of the matroid M. All the bases of a flat B have the same cardinality, called the *rank* of B and denoted $r(B)$. (Characteristic properties of rank functions are given below.) Moreover, the flats of M, when partially ordered by inclusion, form a lattice in which all maximal chains between any two elements have the same length, and the rank of any flat B is the common length of the maximal chains from $\overline{\emptyset}$ to B. In the lattice of flats of any matroid, the infimum of two flats is their intersection, while the supremum is the closure of their union.

Example 3. To illustrate the above concepts, we consider the matroid given by the following affine picture:

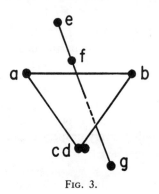

Fig. 3.

We list the flats in decreasing order of their ranks, flats of the same rank being on the same row:

abcdefg
abcd abe abf abg acde acdf acdg bcde bcdf bcdg aefg befg cdefg
ab acd ae af ag bcd be bf bg cde cdf cdg efg
a b cd e f g
∅

The bases of this matroid are the independent 4-sets; in this case the 4-sets that are *not* bases are the sets containing *cd* or *efg*. The family of independent sets is of course the family of subsets of bases.

The characteristic properties of the lattice L of flats of a matroid M are:

L1. L is a finite *point lattice*; i.e., every member of L is the supremum of atoms (elements that cover the zero element of the lattice).

L2. If p and q are atoms of L and $x \in L$, and if $q \leq x \vee p$ but $q \not\leq x$, then $p \leq x \vee q$.

L2 is called the "Mac Lane–Steinitz exchange axiom". To see it in more familiar terms, note that $q \leq x$ if and only if q depends on any set of points whose supremum is x. The axiom then says that if q depends on p and the points of x but not on the points of x alone, then p depends on q and the points of x.

In the converse direction, given a lattice L satisfying L1 and L2 (such a lattice is called a *geometric lattice*), there is a unique geometry whose points are the atoms of L, and whose flats are the sets of points (atoms) under the various members of L.

The elements covered by the maximum element of a geometric lattice (this element corresponds to the flat S of the matroid) are called the *hyperplanes* or *copoints* of the matroid. To illustrate: the projective geometry $PG(n, q)$ can be regarded as the set of all $(q^{n+1} - 1)/(q - 1)$ nonzero vectors of dimension $n + 1$ over the finite field F_q which have the property that the first nonzero entry is 1. These vectors constitute a dependence matroid, which we also denote $PG(n, q)$; it can be checked that the set of vectors whose first entry is 0 forms a hyperplane, and that the remaining vectors form a matroid isomorphic to $AG(n, q)$.

Circuits; Graphical matroids. A third class of examples of matroids is provided by finite graphs. Let S be the set of edges of a graph G, and call a subset A of S independent if it does not contain a circuit, i.e., the set of edges of a polygon of G. (In graph-theoretic terms, the independent sets are edge-sets of "forests" of G.) It is not trivial, but is possible, to check that I1 and I2 are satisfied; the resulting matroid is called the *polygon matroid* of G.

Such matroids are most easily viewed in terms of their dependent sets, viz., sets containing circuits. Since axiom I2 implies that any superset of a dependent set in a matroid is itself dependent, it follows that a matroid is uniquely determined by its family of minimal dependent sets. In the polygon matroid of a graph G, these are precisely the circuits; and in an arbitrary matroid the minimal dependent sets are called *circuits* because of this example.

The characteristic properties of the family $\mathcal{C}$ of circuits of a matroid are:

C1. No member of $\mathcal{C}$ properly contains another,

C2. Given distinct C and $D \in \mathcal{C}$ and $p \in C \cap D$, there is $E \in \mathcal{C}$ contained in $C \cup D - p$.

These can be checked from I1 and I2, and it can be shown also that if $\mathcal{C}$ is such a family, then the family $\mathcal{J}$ of sets not containing members of $\mathcal{C}$ satisfies I1 and I2. In addition, it is evident from graph-theoretical arguments that C2 is satisfied since $(C \cup D) - (C \cap D)$ is in fact an edge-disjoint union of circuits.

Notice that a loop in an arbitrary matroid is just a singleton circuit; in the polygon matroid of a graph, a loop is precisely a loop in the graph-theoretical sense: an edge joining a vertex to itself.

Also, two or more points all depend on each other in an arbitrary matroid (i.e., would be juxtaposed in the affine picture) if every pair of them forms a circuit. In the polygon matroid of a graph, this is a multiple edge: a set of two or more edges joining the same pair of vertices.

Thus, in general, a matroid is a geometry if and only if all its circuits have three or more points; the polygon matroid of a graph is a geometry if and only if the graph has no loops or multiple edges.

Example 4. In the matroid $M(G)$ induced by the graph G shown at right, the circuits are the sets

ab, acde, bcde, acdfg,

bcdfg, efg, h.

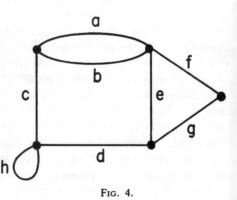

Fig. 4.

The independent sets, or sets containing no circuit, are many in number; the reader can check that the bases, or maximal independent sets, are 19 in number; they all have four edges, and are incident with all vertices of the graph. (This last is a property of any basis of a graphical matroid.)

It is easy to see that the polygon matroid of a graph G is coordinatized over any field by the matrix of 0's, 1's, and -1's whose rows and columns are indexed by the vertices and edges of G, respectively; where for a given orientation (assigning of arrows to the edges) of G a 1 is placed in position (i, j) if the edge j is directed towards the vertex i, a -1 if edge j is directed away from vertex i, and a 0 if i and j are not incident or if j is a loop.

Rank functions; transversal matroids. Before introducing our fourth class of examples, we give the characteristic properties of the rank function r of matroid; it is a nonnegative integer-valued function on subsets of a finite set S satisfying:

R1. $r(\emptyset) = 0$.

R2. If $A \subseteq S$ and $s \in S$, then $r(A) \leq r(A \cup s) \leq r(A) + 1$.

R3. $r(A \cup B) + r(A \cap B) \leq r(A) + r(B)$ for all subsets A, B of S.

R3 is called the axiom of (upper) *semimodularity*.

Given such a function, the sets A for which $r(A) = |A|$ are the independent sets of a matroid whose rank function is r. The bases of a subset A of S are the subsets B of A for which $r(B) = |B| = r(A)$. A loop is a point s with $r(\{s\}) = 0$; a multiple point is a flat A with $r(A) = 1 < |A|$. Thus a matroid is a geometry if and only if $r(A) = |A|$ whenever A is a singleton or a pair.

Now we introduce transversal matroids. Let G be a finite bipartite graph whose vertex set is the disjoint union of sets S and T, and whose edges connect members of S with members of T. (In the popular example of classical matching theory, S is a set of persons, T is a set of consorts, and an edge is drawn between a person and a consort if they are acquainted.) A *matching* of a subset B of S is a one-to-one function f of B into T such that s and $f(s)$ are adjacent in G for every s in B. If such a function exists, we say that B can be *matched* in G.

For each subset A of S, define $r(A)$ to be the size of the largest subset of A that can be matched in G. It is not hard to check R1,

R2, and R3; the resulting matroid is called the *transversal matroid induced by G on S*. We say that G is a *presentation* of this matroid, or that the matroid is *presented* by G.

Of course, the independent sets of this matroid are simply the sets that can be matched. It was R. Rado who noticed in 1942 [34] that these sets do constitute the independent sets of a matroid; but it was in 1965 that the subject was brought to general attention by Edmonds and Fulkerson [23]. (The latter paper, incidentally, appears in a volume (69B) of the Journal of Research of the U.S. National Bureau of Standards that is remarkable on account of its interest to workers in this field. In addition to the paper mentioned, it contains important papers by Tutte [40] and Crapo [12] and a number of other papers of combinatorial interest.)

In the transversal matroid on S induced by G, a loop is a "person who knows no consort", and a multiple point is a set of "persons who collectively know but one consort."

Example 5. Consider the bipartite graph shown.

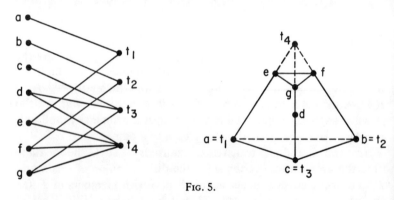

Fig. 5.

The reader can check that the bases of the transversal matroid are exactly those of the affine matroid in Example 2 above. We show also an affine picture of this same matroid; we have completed this picture to a tetrahedron to illustrate the correspondence between transversal and "free-simplicial" matroids made in [10]. Informally, a set S of points in affine space is *free-simplicial* if there is a simplex (whose vertices may or may not be in S) such that the points of S are in the faces of the simplex, and such that a point in the interior of a face is in free position with respect to every other set of points on the face. It then turns out that a matroid is transversal if and only if it is free-simplicial. The correspondence is as suggested by our illustration; the vertices of the simplex correspond to the members of T, and a member s of S is related to the vertices of the smallest face of the simplex in which s lies.

Crapo and Rota [14] have noted another class of matroids arising from simplicial complexes: these generalize in a striking way the graphical matroids $M(K_n)$ of the complete graphs. This generalization runs as follows: For any set E in a matroid M let $\beta_1(E)$ be the maximum size of a set $\{C_1, \ldots, C_k\}$ of circuits contained in E with the property that $C_i \not\subseteq C_1 \cup \cdots \cup C_{i-1}$, $i = 1, \ldots, k$. It can be checked that the rank of E is $r(E) = |E| - \beta_1(E)$.

Now in the matroid $M(K_n)$, let $\beta_0(E)$ be one less than the number of connected components of the subgraph of K_n consisting of all n vertices and the edges of E. Then $r(E) = n - 1 - \beta_0(E)$.

In the simplicial complex whose one-simplices, $T_1(E)$, are the vertices of K_n and whose two-simplices, $T_2(E)$, are the edges in E, $\beta_0(E)$ is the dimension of the zero reduced simplicial homology group, the zero Betti number; while $\beta_1(E)$ is the first Betti number.

These concepts generalize to higher dimensional simplicial complexes. Indeed, let the *simplicial geometry* $M(T_{n,k})$ be the matroid whose points correspond to all k-subsets T_k of a set T with n elements, and associate with every subset A of $M(T_{n,k})$ the simplicial complex whose i-simplices are the i-subsets of T ($0 \le i \le k - 1$) and whose k-simplices are the elements of A. Then when the associated Betti numbers are computed, $\beta_i = 0$ for all $i \ne k - 1, k - 2$; while $r(A) = \binom{n-1}{k-1} - \beta_{k-2}(A) = |A| - \beta_{k-1}(A)$ is a matroid rank function. Here, $M(T_{n,2}) = M(K_n)$.

Linear and algebraic dependence and a historical note. The characteristic properties of the relation of dependence (of an element on a subset) in a matroid are:

D1. If s is a member of A, then s depends on A.

D2. If s depends on $A \cup t$ but not on A, then t depends on $A \cup s$.

D3. If s depends on A and every member of A depends on B, then s depends on B.

(D2 will be recognized as the Mac Lane–Steinitz exchange axiom.) As with all the other axiom systems, a relation satisfying these gives rise to a matroid in the obvious fashion, and the dependence relation of any matroid satisfies these.

For our final example of matroids arising in classical mathematics, we consider a field K and an extension field L of K, and we say that $A = \{a_1, \ldots, a_r\} \subseteq L$ is independent if $a_1, \ldots, a_r$ are algebraically independent over K, i.e., if there is no polynomial in r variables with coefficients in K of which $(a_1, \ldots, a_r)$ is a root. If S is any finite subset of L, then the algebraically independent subsets of S are the independent sets of a matroid. The rank of this *algebraic dependence matroid* is called the *transcendence degree* of (the algebraic closure of) S over K.

In the 1937 edition of *Moderne Algebra* [42], B. L. van der Waerden treated both linear dependence in a vector space over a field and algebraic dependence in an extension of a field from the viewpoint of axioms D1, D2, and D3. In both treatments he proved these axioms first and then derived other properties from them using "coordinate-free" arguments, i.e., without making use of the definition of the dependence relation in question.

But in the 1930 edition of the same book, the treatment of these subjects, while including essentially the same results, did not recognize the fact that there is a set of abstract axioms that serves for both types of dependence.

What happened between these two editions was an explosion of ideas at Harvard University in the years prior to 1935, culminating in the appearance in volumes 57 and 58 of the *American Journal of Mathematics* of three pioneering papers by Hassler Whitney [45], Saunders Mac Lane [33], and Garrett Birkhoff [2]. (Because these papers and their authors were so close to each other, it seems pointless to try to assign precedence; nevertheless, Birkhoff mentioned each of the other two papers, Mac Lane mentioned Whitney's, and Whitney mentioned neither of the others.)

In Whitney's paper, entitled "On the abstract properties of linear dependence", he coined the term "matroid" and gave the equivalent axiomatizations in terms of rank function, independent sets, circuits, and bases; he also discussed the notions of separability, duality, and coordinatizability (which we will consider below). In addition to giving the example of dependence matroids, he observed that the edges of a graph form a matroid as described above, and also showed that such a matroid is always a dependence matroid over (i.e., is *representable* or *coordinatizable* over) the field of two elements.

In his paper, Mac Lane connected the notions of linear dependence, geometric lattices, and projective and affine geometry, introduced what we have called "affine pictures", and also observed that algebraic dependence over a field satisfies the same axioms as does linear dependence. And in Birkhoff's paper matroids were defined for the first time in terms of their lattices of flats; connections with projective geometry were also discussed.

II. CONSTRUCTIONS

This section contains four types of geometrical constructions: ways of combining and transforming matroids to produce new matroids.

Direct sum and matroid union. Let $M_1 = (S_1, \mathcal{J}_1)$ and $M_2 = (S_2, \mathcal{J}_2)$ be matroids on disjoint sets S_1 and S_2, and define the family $\mathcal{J}$ of subsets of $S = S_1 \cup S_2$ by saying that A is in $\mathcal{J}$ if $A \cap S_i$ is in $\mathcal{J}_i$, $i = 1, 2$. Then $(S, \mathcal{J})$ is a matroid, called the direct sum of M_1 and M_2, and denoted $M_1 \oplus M_2$.

A basis of $M_1 \oplus M_2$ is the union of a basis of M_1 and a basis of M_2; the circuits of $M_1 \oplus M_2$ are the circuits of the respective geometries. The rank function r of the direct sum is given by $r(A) = r_1(A \cap S_1) + r_2(A \cap S_2)$, where r_i is the rank function of M_i; and the lattice L of flats of the direct sum is the cartesian product of the lattices L_1 and L_2 of M_1 and M_2, with the partial order $(A, B) \leq (C, D)$ if $A \leq B$ in L_1 and $C \leq D$ in L_2.

For affine matroids, $M_1 \oplus M_2$ is pictured by placing pictures representing M_1 and M_2 in general position relative to each other in a space of sufficiently high dimension (one greater than the sum of the dimensions of the spaces for M_1 and M_2). And if M_1 and M_2

are dependence matroids over the same field, then a matrix representation for their direct sum is

$$\begin{pmatrix} D_1 & 0 \\ 0 & D_2 \end{pmatrix}$$

where D_i is a matrix representing M_i, $i = 1, 2$.

For graphical and transversal matroids, $M_1 \oplus M_2$ corresponds to a graph that is the disjoint union (or wedge product for graphical matroids) of graphs whose matroids are M_1 and M_2.

A matroid is *connected* if it is not the direct sum of nonempty matroids; indeed, one principal use of the notion of direct sum is the reduction of much of the study of matroids to the study of connected matroids.

Another use of the notion is in the area of coordinatizability. It is not difficult to check that if $M_1 \oplus M_2$ is coordinatizable over a field K, then so are M_1 and M_2, and conversely. (In fact, a direct summand is a particular kind of "minor" (see below), and it is easy to show that a matroid is coordinatizable over K if and only if all its minors are.) Thus if we take two fields K_1 and K_2 and a matroid M_i that is not coordinatizable over K_i ($i = 1, 2$), then $M_1 \oplus M_2$ is not coordinatizable over any field at all. Such phenomena will be discussed in more detail below.

As a generalization of direct sum (as well as of transversal geometries) we may define the matroid union $M(S, \mathcal{I}) = M_1 \vee M_2$ of two matroids, $M_1(S, \mathcal{I}_1)$ and $M_2(S, \mathcal{I}_2)$, on the same set by $\mathcal{I} = \{I_1 \cup I_2 | I_1 \in \mathcal{I}_1, I_2 \in \mathcal{I}_2\}$. With this definition, the rank function is given by $r(A) = \min_{B \subseteq A} (r_1(B) + r_2(B) + |A - B|)$. Then transversal matroids are precisely successive unions of rank-one matroids, and we may get the direct sum $M_1(S_1) \oplus M_2(S_2)$ by adding loops to augment S_1 and S_2 to a common set. The reader is encouraged to check that $M_1 \vee M_2$ is coordinatizable over F whenever M_1 and M_2 are coordinatizable if and only if F is infinite, in which case the representation can be effected by placing the matrix representation for M_1 on top of that for M_2 after the columns of M_2 have been multiplied by appropriate scalars. To realize $M_1 \vee M_2$ affinely (over a suitably large field), one places the matroids M_1 and M_2 in general relative position, and then puts $p \in M_1 \vee M_2$ freely on the line between $p \in M_1$ and $p \in M_2$.

MATROIDS AND COMBINATORIAL GEOMETRIES   193

In [50] matroid union is used to construct a polynomial-time algorithm to find the largest common independent set of two matroids on the same set. The determination of the cardinality of such a set has many applications. For example, in [52] it is shown that determining the existence of a Hamiltonian circuit in a graph G (for which no polynomial algorithm is known to exist) is equivalent to ascertaining whether three matroids associated with G have a common basis. The difference between common basis computation for two and for three matroids seems to be that in the former case the computation can be made within one associated matroid while in the latter case no one such matroid has been constructed in general.

The inference that when an associated matroid exists, efficient algorithms can be found is further substantiated by the "greedy" property of matroid bases as illustrated on page 203.

Deletion (subgeometry). Let $M = (S, \mathcal{I})$ be a matroid and T a subset of S. Define $\mathcal{I}_T$ to be the family of intersections of members of $\mathcal{I}$ with T, i.e., the family of members of $\mathcal{I}$ that are subsets of T. Then $(T, \mathcal{I}_T)$ is a matroid, called the *deletion* of $S - T$ from M, or the *subgeometry* of M on T. We use one of two notations for this deletion: either $M(T)$ or $M - T'$, where T' denotes the subset $S - T$.

It can be checked that the circuits of $M(T)$ are just the circuits of M that are subsets of T, and that the flats of $M(T)$ are the intersections of T with the flats of M.

Given any representation of a matroid M (by means of a matrix, affine picture, graph, or transversal structure), a representation of $M(T)$ can be effected by simply deleting all elements of $S - T$ from the representation while preserving all other structural relationships.

A trivial but useful fact about deletions is that if $T' = S - T = \{t_1, \ldots, t_r\}$, then $M(T) = (\ldots((M - t_1) - t_2) - \ldots) - t_r$, the single-point deletions being carried out in any order whatever.

Contractions. If $M = (S, \mathcal{I})$ is a matroid, T is a subset of S, and $T' = S - T$, then we define $\mathcal{I}/T'$ to be the family of subsets of T whose unions with some fixed basis of T' are in $\mathcal{I}$. Then $(T, \mathcal{I}/T')$ is a matroid denoted M/T', called the *contraction* of M by T'. The flats of M/T' are simply the flats of M containing T', with

T' subsequently deleted; the lattice of M/T' is isomorphic to the interval $[\overline{T'}, S]$ of the lattice of M.

As with deletions, contractions can (and should) be carried out one point at a time in any order: if $T' = \{t_1, \ldots, t_r\}$, then $M/T' = (\ldots((M/t_1)/t_2)/\ldots)/t_r$.

To realize a one-point contraction M/p of an affine matroid, we merely picture the projection from p of the points of $S - p$ onto a hyperplane of affine space that is not parallel to any of the lines of M through p. For example, if M is pictured as shown below, then M/a has the picture shown at right.

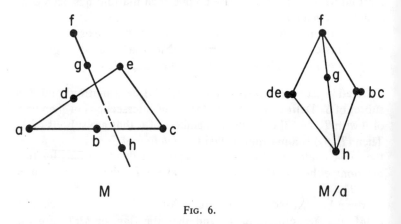

Fig. 6.

(The fact that in an affine space over a finite field, such a freely placed hyperplane might not exist can be used to prove that contractions of affine geometries over finite fields need not be affine.)

Let M be the polygon matroid of the graph G, and suppose p is an edge of G that is not a loop. Then M/p is the polygon matroid of the graph obtained from G by "contracting p to a point"; that is, by identifying the two vertices incident with p (removing the induced loop p) while preserving all incidences of other edges with these vertices.

Let M be the dependence matroid of the columns of the matrix D, and suppose p is not a loop, i.e., is a nonzero column of D. Then M/p is the dependence matroid of the matrix D/p obtained from D by using elementary row operations to reduce the number of nonzero entries in column p to one, then deleting the row of this nonzero entry and the column p. Note that this induces a linear transformation from the column space of D to that of D/p. The contraction of a transversal geometry is not, however, transversal in general. For example, a quadrilateral with three multiple points can be imbedded freely in a 2-simplex; but when the fourth point is contracted, the resulting matroid cannot be freely embedded in a 1-simplex.

We remark that any subgeometry of a geometry must be a geometry; but contractions can produce loops and multiple points. (The reader can check that loops are produced whenever T' is not a flat of M.)

Any combination of deletions and contractions is called a *minor* of a matroid; it is easily checked that deletions and contractions (of disjoint sets) commute, so that, for example, $((M/A) - B)/C = ((M - B)/(A \cup C))$. In particular, any minor can be effected as the result of a sequence of one-point deletions, followed by a sequence of one-point contractions (or vice versa).

Dual matroids. As with circuits, independent sets, and flats, the bases of a matroid may also be characterized in an intrinsic manner: they form a nonempty family of subsets such that:

B1. No basis properly contains another.

B2. If B_1 and B_2 are bases and $b_1 \in B_1$, then there is a point $b_2 \in B_2$ such that $(B_1 - b_1) \cup b_2$ is a basis.

The notion of a basis obviously suggests its specialization in dependence matroids, but bases also play significant roles in transversal matroids where they are subsets which admit maximal matchings, and in connected graphs where they are the spanning trees.

It was a discovery of Whitney that the *basis exchange axiom* B2 (as well, of course, as B1) is also satisfied by the *complements* of bases. The resulting matroid M^* (on the same set S as M) is termed the *(Whitney) dual* of M (or occasionally the *orthogonal matroid* to M). Thus $(M^*)^* = M$.

To illustrate an instance of this idea of the basis complement, let P be a convex polyhedron (with labeled faces) in three dimensions. Consider all the ways of cutting along a subset of edges E_c of P and unfolding along the other edges E_f to produce a planar (connected) model. Thus, e.g., a cubical die may be unfolded as:

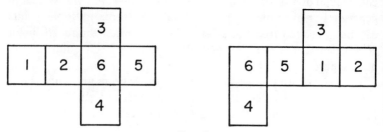

Fig. 7.

It is easy to see that a subset of edges qualifies as a subset E_c if and only if it meets every vertex and contains no circuits (i.e., is a spanning tree of the associated planar graph, or basis of its geometry). On the other hand, a subset of edges will be the subset E_f left uncut (and used for the unfolding) if and only if when a vertex is placed in each region and edges are joined across two faces which are both incident with an E_f-edge, the resulting set of edges forms a "face spanning tree" (spanning tree of the planar dual graph of P). By extension, we can use the dual of a planar graph to show that the dual of such a matroid is graphical. And in fact the converse is also true: the dual of a nonplanar graphical geometry is not graphic.

Every geometric concept we have discussed can be characterized in the dual. For example, circuits of G are complements of hyperplanes in G^*, the rank function r^* of G^* is given in terms of the rank function r of G by the formula $r^*(A) = r(S - A) + |A| - r(S)$. Using the above rank function, it is a direct consequence of the Alexander duality theory of Betti numbers that for simplicial geometries, $M^*(T_{n,k}) = M(T_{n,n-k})$. The dual of a loop is an *isthmus*, a point in no circuit (and thus in every basis).

For a dependence matroid $M(D)$ of rank r and cardinality n the dual matroid has two linear interpretations. First, if D^* is a subset of rows which span the subspace orthogonal to the row space V_D of D, then $M^*(D) = M(D^*)$. Thus, if r columns of D form an identity submatrix I_r so that $D = [I_r\, D\,']$; then $D^* = [(-D\,')^T\, I_{n-r}]$

(with the same column index and where A^T denotes matrix transpose) coordinatizes $M^*(D)$. Thus in Example 1, where I_2 consists of the third and fourth columns of D_2,

$$D_2^* = \begin{bmatrix} 1 & 0 & 0 & 0 & 0 \\ 0 & 1 & 0 & -1 & 0 \\ 0 & 0 & -1 & -1 & 1 \end{bmatrix}$$

A second representation of $M^*(D)$ is as the support matroid of the rowspace V_D. The *support* of a row vector consists of those columns in which it has nonzero entries. A row vector of minimal support in a subspace V_D is a nonzero vector in V_D whose support properly contains the support of no other nonzero vector of V_D. The *support matroid* $C(M)$ on S has for its circuits the subsets C of S which are the minimal supports of vectors in V_D. Vectors of minimal support of D_2 in Example 1 are $r_1 = (0, 0, 1, 0, 1)$; $r_2 = (0, 1, 0, 1, 1)$, and $r_1 - r_2 = (0, -1, 1, -1, 0)$. Thus, the circuits of $M^*(D)$ are *ce, bde,* and *bcd*.

Over infinite fields, duals of affine geometries are also realizable as subsets of affine space. For instance, the dual of the matroid in Example 2 is pictured in the Euclidean plane as follows:

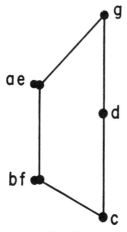

FIG. 8.

Duals of transversal matroids are not necessarily transversal, the above matroid providing an example of one which can not be imbedded freely on a simplex. However, work of Brown, Dowling, and Kelly [18] has characterized duals of transversal geometries in terms of ·the categorical structure of matroids and certain morphisms called strong maps discussed in [14]. Another model for the class of dual transversal matroids is provided by *strict gammoids* [52], where S is the set of vertices of a directed graph G, S' is a distinguished subset of S, and vertices are independent if they can be linked into S' by edge disjoint (directed) paths.

Other areas in which matroid duality plays an important role are in linear programming and in network theory where the space of edge-flows in a directed graph is dual to the space of tensions (potentials).

A geometry is a *cube* if it is isomorphic to a geometry of rank four on the eight points $\{a, b, c, d, e, f, g, h\}$ which includes among its circuits $\{a, b, c, d\}$, $\{c, d, e, f\}$, $\{e, f, g, h\}$ and $\{g, h, a, b\}$. Examples of cubes include the eight vectors in an affine space consisting of only zeros and ones (and in particular AG(3, 2)). The reader is invited to check some of the ideas presented in this section by ascertaining which cubes are *self-dual* (i.e., satisfy $C \simeq C^*$).

Matroid constructions are related to each other by certain commutativity relations. For example, if A and B are disjoint:

$$((G/A) - B)^* = (G^* - A)/B,$$

$$(G_1 \oplus G_2)^* = G_1^* \oplus G_2^*,$$

$$((G_1 - A)/B) \oplus G_2 = ((G_1 \oplus G_2) - A)/B, \text{ etc.}$$

III. REPRESENTATIONS OF GEOMETRIES

As in other branches of mathematics, many of the interesting problems in the theory of matroids concern the general question of an intrinsic characterization of those geometries which can be represented by more classical combinatorial structures (as a subset of vectors in a vector space, edges in a graph, points in an affine space, etc.). These problems are all difficult and we will now mention some of the progress which has been made along these lines.

For a class $\mathcal{C}$ of geometries such as those represented by graphs or dependence geometries over a fixed field F (or fixed characteristic p), we have seen that minors of such geometries are themselves representable. This property may be restated as follows: There is a minimal (but perhaps infinite) class $\mathcal{O}$ of matroids such that a geometry is in $\mathcal{C}$ if and only if it has no minor isomorphic to a member of $\mathcal{O}$. This is called a *characterization of $\mathcal{C}$ by the forbidden minors* (or *obstructions*) $\mathcal{O}$. As illustrations of this idea the reader should convince himself that a loop is the obstruction characterizing the class of boolean algebras; and that the direct sum of a loop and an isthmus is the obstruction for free geometries.

The first interesting characterization by forbidden minors dates back to the origins of the theory: Whitney [45] proved that a matroid M is *binary*, i.e., representable over the field with two elements, if and only if M does not have a four-point line $L_4 = F^{4,2}$ as a minor.

A much more difficult characterization is that of Tutte [40] for *unimodular* matroids (those which can be represented over the rational field as a dependence geometry in which every subdeterminant (and entry) is zero, one, or negative one). Tutte showed that these matroids are precisely those binary matroids which are also coordinatized over a field (equivalently every field) of characteristic other than two. The excluded minors for such geometries are L_4; the Fano plane, $F = PG(2, 2)$ as in Figure 9 (in which the curve denotes a line); and its dual F^* (the geometry resulting when one point is deleted from the cube $AG(3, 2)$).

In the same article Tutte showed that graphical geometries were unimodular geometries which in addition did not have the duals of the matroids arising from the two Kuratowski graphs (K_5, the complete five graph; and $K_{3,3}$, the "three persons to the three wells" graph).

Two other obstruction theorems are known: Reid, Seymour, and Bixby [3], independently showed that a geometry is coordinatizable over F_3 if and only if it does not contain L_5, the five-point line;

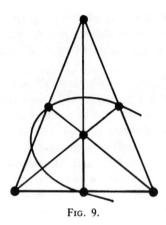

Fig. 9.

L_5^*; the Fano plane F; or F^*. It is perhaps not surprising that an obstruction theory for linear representability has been completely developed only for binary, ternary, and unimodular matroids. It is precisely these three cases in which coordinatizations are unique in that any two coordinatizations of the same matroid are projectively equivalent [49].

We have seen that transversal matroids do not form a class which is closed under the taking of arbitrary minors. However, if we augment the class of transversal matroids to include their minors, results of Duffin [22], Brylawski [5, 10] and others have shown that the class of binary geometries which are minors of transversal geometries are exactly those representable by graphs which are series-parallel networks: those matroids which do not have as minors L_4 and $M(K_4)$, the matroid represented by the complete four graph.

Forbidden minors which characterize the class of geometries coordinatizable over some field lead to interesting (but by no means totally classified) geometries: they include the planar "failed" Pappus and Desargues configurations wherein incidences forced by the theory of coordinate geometry are not forced by weaker combinatorial conditions. Other minimal counterexamples to coordinatization include cubes which contain the plane $\{a, b, e, f\}$ but do not contain its complementary plane. In fact, these cubes cannot be represented as algebraic dependence matroids [31]. Another minimal counterexample to coordinatizability has been constructed by Lindström, who adjoins points to the Fano plane (Figure 9) giving another subgeometry isomorphic to the Fano plane but where the points on the curved line are independent. He shows, however, that this can be coordinatized algebraically over $F_2[x, y, z]$.

A coordinatization problem related to the obstruction problem is that of determining which sets of integers are *characteristic sets*. The characteristic set $\text{Char}(G)$ of a geometry G is defined by saying that $p \in \text{Char}(G)$ if and only if G is coordinatizable over some field of characteristic p. The classical theory of projective geometries shows that $\text{Char}(PG(n, p)) = \{p\}$ (for $n \geq 2$) so that all singletons form characteristic sets. (An affine instance of the fact that $0 \notin \text{Char}(PG(2, p))$ is the theorem of Gallai [30]: finite planes in which every line has three or more points cannot be drawn in E^2 with straight lines; this explains, for example, our curved line in the picture of the Fano plane.)

We have also given several examples of geometries with empty characteristic sets.

In addition, any set containing 0 and all but finitely many primes is a characteristic set. To see this, first note that a dependence geometry $M([I \; D'])$ over one field is isomorphic to a dependence geometry $M([I \; D''])$ over another field if and only if corresponding subdeterminants of the submatrices D' and D'' vanish or fail to vanish simultaneously. Then, following the example of R. Reid, let p be a prime, ω a complex primitive pth root of unity, and $P(k, \omega) = 1 + \omega + \omega^2 + \cdots + \omega^k$; and let M be the dependence geometry (over the complex numbers) $M([I \; D_p'])$ where

$$D_p' = \begin{bmatrix} 0 & 0 & 0 & & 0 & 1 & 1 & 1 & & 1 & 1 \\ 1 & 1 & 1 & \cdots & 1 & 1 & 1 & 1 & \cdots & 1 & 1-\omega \\ 1 & P(1, \omega) & P(2, \omega) & & P(p-2, \omega) & 0 & 1 & P(1, \omega) & & P(p-2, \omega) & 1 \end{bmatrix}.$$

The reader should convince himself by checking determinants that for a prime p', M can be represented over a field of characteristic p' if and only if it can be represented over the extension field

$F_{p'}(\omega')$, where ω' is a primitive root of the polynomial $P(p-1, x) = (x^p - 1)/(x - 1)$ ($\omega' \neq 1$). But this can be accomplished for all characteristics p' except when $p' = p$, in which case $P(p-1, x) = (x-1)^{p-1}$. Thus if $\mathcal{P}$ is the set of all primes (and 0), $\text{Char}(D_p') = \mathcal{P} - p$. Using direct sums, we now see that all cofinite sets of primes can exist as characteristic sets.

Two theorems of P. Vamos and R. Rado guarantee that the following three statements are equivalent: $0 \in \text{Char}(G)$; $\text{Char}(G)$ is infinite; $\text{Char}(G)$ is cofinite. Therefore, all characteristic sets are either finite or cofinite. In addition to the above examples, the only other case decided is that $\{1103, 2089\}$ is a characteristic set!

Another way of characterizing the geometries of various classes is by noting how the axiom systems of the various ways to describe a geometry (e.g., for circuits, bases, etc.) can be strengthened when the geometry has a certain representation.

For example, let M be a square nonsingular matrix and apply the Laplace determinant expansion theorem to a subset A of its columns. Then since $\det M \neq 0$, there exist two nonvanishing complementary square subdeterminants of M, the first with columns indexed by A and the second indexed by the complementary set of columns. Couched in geometric terms, this means that if B_1 and B_2 are two bases of a vector space with A_1 a subset of B_1, and both are expressed in terms of B_2 coordinates, resulting in the matrices M and I respectively, then there is a subset A_2 of B_2 such that $(B_1 - A_1) \cup A_2$ and $(B_2 - A_2) \cup A_1$ are both bases (corresponding to nonsingular matrices). This seems quite a strengthening of the basis exchange axiom B2. (For example, a weaker strengthening is: for all $b_1 \in B_1$, there is a $b_2 \in B_2$ such that *both* $(B_1 - b_1) \cup b_2$ and $(B_2 - b_2) \cup b_1$ are bases.) Does this subset exchange help to characterize those geometries coordinatizable over a field? The answer is no, since it is shown in [26], [50], and [7] that all combinatorial geometries enjoy this seemingly stronger axiom. The question remains: Are there other determinantal identities which when reformulated combinatorially for bases of a geometry either have general combinatorial proofs (thus simultaneously proving theorems for maximal matchings, transcendence bases, etc.) or do not hold in general (thereby supplying necessary conditions for coordinatizability)?

Another stronger basis exchange axiom which holds in general is that there exists a bijection $f: B_1 \to B_2$ such that for all $b \in B_1$, $(B_1 - b) \cup f(b)$ is a basis. As noted by Brualdi [4], it is not true in general that there is a bijection f such that both $(B_1 - b) \cup f(b)$ and $(B_2 - f(b)) \cup b$ are bases (the reader may want to check this for himself with the matroid $M(K_4)$); but this property is satisfied for all transversal matroids and their minors). Thus we have found a necessary (but alas not sufficient) condition for minors of transversal matroids.

Again, binary matroids have afforded the best results along the above lines. For example, binary geometries are characterized by the fact that they allow a strengthening of B2 so that the number of elements of B_2 which can be exchanged with a fixed element b of B_1 resulting in two new bases is always odd. Another characteristic strengthening is that for binary geometries, the circuit elimination axiom C2 becomes: $(C_1 \cup C_2) - (C_1 \cap C_2)$ is a disjoint union of circuits (as we saw was true for graphical geometries).

An axiomatization for bases which shows how matroid structure improves algorithmic complexity follows from the "greedy algorithm": a collection of subsets of S is the family of bases of a matroid if and only if for every assignment of real-number weights to the points of S, the subset with the greatest weight (sum) is the subset greatest in the lexicographic ordering induced by the assignment. Thus, in the matroid case, determining the maximally weighted basis can be found very quickly as no backtracking need be done.

A systematic study of the relationship between bases and the classical theory of determinants has been carried out by Neil White via the *bracket ring* or *ring of syzygies* of a combinatorial geometry [43].

Given a matroid M of rank k on a set S, the bracket ring $B(M)$ is obtained by taking the quotient of the free ring on the set of all ordered k-tuples (*brackets*), $[x_1, \ldots, x_k]$, of S by the two-sided ideal generated by the following relations (or *syzygies*) for all $x_1, \ldots, x_k, y_1, \ldots, y_k \in S$:

1. $[x_1, \ldots, x_k]$ if $\{x_1, \ldots, x_k\}$ is a dependent subset of M or contains repeated elements.

2. $[x_1, \ldots, x_k] - (\text{sgn } \sigma)[x_{\sigma(1)}, \ldots, x_{\sigma(k)}]$ for any permutation σ of $\{1, \ldots, k\}$.

3. $[x_1, \ldots, x_k][y_1, \ldots, y_k] - \sum_{i=1}^{k} [x_1, \ldots, x_{i-1}, y_1, x_{i+1}, \ldots, x_k]$ $[x_i, y_2, \ldots, y_k]$.

Note that the second and third conditions are reminiscent of the quadratic relations in the Grassmannian for the point corresponding to a particular $(k - 1)$-plane in $PG(|S| - 1, F)$.

An arbitrary matroid can be represented in a certain sense in a module over its bracket ring. This representation is not strictly a coordinatization but reflects many features of the structure of the matroid. The reader may easily work through the details showing that if M is a dependence matroid $M(D)$ where M is a $k \times n$ matrix over F, then $[x_1, \ldots, x_k] \mapsto \det(c_1, \ldots, c_k)$ defines a homomorphism $B(M) \to F$ (where c_i is the column of M corresponding to x_i). Conversely, for an arbitrary matroid M, if $h: B(M) \to F$ is any ring homomorphism such that $h([x_1, \ldots, x_k]) \neq 0$ whenever $\{x_1, \ldots, x_k\}$ is a basis of M, then a matrix D may be constructed so that $M = M(D)$. White has characterized binary geometries in terms of the structure of their bracket rings. He has shown that a binary matroid M is unimodular if and only if the radical of $B(M)$ is a prime ideal, and it is conjectured that this holds if and only if $B(M)$ is an integral domain.

IV. RESEARCH IN GEOMETRIC LATTICES

What has suggested the direction for much of the current research into the structure of geometric lattices is that these lattices are generalizations of the finite *partition lattice* Π_n, the geometric lattice associated with the circuit geometry of the complete graph K_n. Three properties of partition lattices are especially significant. In the first place, they are supersolvable in the sense of [37]; that is, they have saturated chains of modular elements (x is modular if $r(x) + r(y) = r(x \vee y) + r(x \wedge y)$ for all $y \in L$). In the polygon matroid of K_n, such a chain is an increasing chain of cliques (i.e., complete subgraphs on the vertices $v_1, \ldots, v_j$ for $j =$

1, 2, ..., n). The existence of such a chain provides a facile computation of the characteristic polynomial of the lattice.

Indeed, let $\mu(x, y)$, the *Möbius function* of a lattice as defined in [35], be defined by the recursion $\mu(x, y) = 0$ if $x \not\leq y$; $\mu(x, x) = 1$; and for $x < y$, $\mu(x, y) = -\sum_{x \leq z < y} \mu(x, z)$. Then the characteristic polynomial (which we will study further in the following section) of a loopless matroid is defined by

$$\chi(M) = \chi(M, \lambda) = \sum_{A \in L} \mu(0, A)\lambda^{r(M)-r(A)},$$

the sum being taken over the flats $\{A\}$ in the geometric lattice L of M. (If M has a loop then $\chi(M) = 0$.) Then, if L is supersolvable with a saturated chain of modular flats $0 < A_1 < A_2 < \cdots < A_n = S$, $\chi(M) = \prod_{i=1}^{n} (\lambda - \alpha_i)$ where a_i equals the number of atoms in A_i but not in A_{i-1}.

These observations have led to other studies of modular elements in geometric lattices [36], [9].

A second way that partition lattices have motivated study into geometric lattices is that an important structure theorem in the theory of lattices [16] states that every lattice is a sublattice of the lattice of partitions of some (perhaps infinite) set. A famous problem posed by Dilworth and Whitman asks if every finite lattice is a sublattice of Π_n for a sufficiently large n. In attempting to answer this question, Dilworth [16] proves that every finite lattice is a sublattice of a geometric lattice. A sketch of his proof is outlined as follows:

For a finite lattice L', we define recursively (from the top), $\sigma(1) = 0$; $\sigma(q) = \sigma(q') + 1$ if q is (a meet-irreducible element) uniquely covered by q'; and otherwise $\sigma(a) = \max_{x \wedge y = a} (\sigma(x) + \sigma(y) - \sigma(x \vee y))$. Then $r'(a) = \sigma(0) - \sigma(a)$ is an (upper) semimodular strictly increasing function on L' which is 0 on the zero of the lattice. For all join-irreducible elements q (i.e., q covers a unique element q' of L'), let S_q be a set with $r'(q) - r'(q')$ elements and let $S = \{S_q | q \text{ is a join-irreducible element of } L'\}$. Let $A_a = \cup \{S_q | q \leq a\}$. Then the elements $\{A_a | a \in L'\}$ form a closure system $\mathcal{C}$ (i.e., a system containing S and closed under intersections).

If we define $r(A) = \min_{C \in \mathcal{C}} (r'(C) + |A - C|)$, then r is a matroid rank function for a matroid M with geometric lattice L such that members of $\mathcal{C}$ are flats of M and form a sublattice of L isomorphic to L'.

To illustrate these ideas, we note that when the geometric lattice associated with the free geometry $F^{4,3}$ is inverted, giving the lattice L', L' is not geometric, but that the above methods embed it in the lattice L of the affine geometry generated by the vertices of a truncated tetrahedron in Euclidean space:

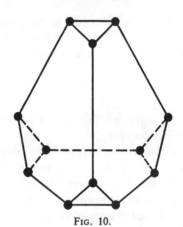

Fig. 10.

L' corresponds to the six original tetrahedron edges and four hexagonal faces respectively. The above embedding theorem makes the "sublattice of Π_n problem" equivalent to one in geometries—can every finite lattice of geometric flats be represented by a subfamily of partitions of a finite set S with intersection of flats corresponding to common refinement, and with supremum likewise being represented by the supremum operation in $\Pi_{|S|}$?

P. Pudlák and J. Tuma of Prague, using techniques from universal algebra, have proved Whitman's conjecture by embedding any lattice L into a partition lattice on a finite (but enormous) set S. Using their construction, embedding a ten-element lattice L requires that S contain billions of points. Because of the obvious applications of this theory to computer storage and manipulation, embeddings into much smaller partition lattices are desired. Since Dilworth's construction is quite efficient in this regard, one feels that a theory for embedding geometric lattices in small partition lattices should be explored (see, e.g., recent work by R. Peele).

A third area for research comes from the geometric generalization of the Stirling numbers. The *Stirling numbers of the second kind* $S(n, k)$ recur frequently in combinatorial arguments (for example, $S(n, k)$ counts the number of functions from a set with n elements onto a (delabeled) set with k elements and also serves as the coefficient of the falling factorial $x(x - 1) \cdots (x - k + 1)$ in the expansion of x^n). For our purposes, they gauge the width of the lattice Π_n. In particular, $S(n, k)$ is the number of flats of rank $(n - k)$ in the geometric lattice Π_n ($k = 1, 2, \ldots, n$). Lieb [32] has proved that these numbers are logarithmically concave: $(S(n, k))^2 \geq S(n, k - 1)S(n, k + 1)$, and therefore *unimodal*: $S(n, j) \geq \min(S(n, i), S(n, k))$ for all $i \leq j \leq k$. We may now check analogous properties for an arbitrary geometric lattice L, in which the *Whitney number of the second kind* $W(k)$ is the number of elements of rank k. Thus $S(n, k) = W(n - k)$ in π_n.

Other examples of Whitney numbers which have been proved to be logarithmically concave are for Boolean algebras where they are binomial coefficients; and for $PG(n - 1, q)$, where the Whitney numbers correspond to *Gaussian coefficients*

$$W(k) = \begin{bmatrix} n \\ k \end{bmatrix}_q = \frac{q^n - 1}{q - 1} \cdot \frac{q^{n-1} - 1}{q^2 - 1} \cdot \ldots \cdot \frac{q^{n+1-k} - 1}{q^k - 1}.$$

Only partial results are known about the possibilities for the general sequence $1 = W(0), W(1), \ldots, W(n) = 1$. In particular, $W(1) \leq W(n - 1)$ [1, 25], and in fact $\sum_{i=0}^{k} W(i) \leq \sum_{j=n-k}^{n} W(j)$ with equality holding for some $k \in [1, n - 1]$ if and only if it holds for all k; equivalently, if and only if every element of L is modular.

The proof of this latter statement, found in [21], makes elegant use of the property that in a geometric lattice, $\mu(x, 1)$ (and in fact $\mu(x, y)$) is never zero, thus allowing a Möbius inversion [35] which in turn along with semimodularity shows that a vector space of dimension $\sum_{i=0}^{k} W(i)$ is spanned by a certain subset of $\sum_{j=n-k}^{n} W(j)$ vectors.

Also, if $W(1) = n + m$, then

$$\binom{n}{k} + \binom{n-2}{k-1} m \leq W(k) \leq \binom{n+m}{k} \quad [20].$$

Very little else is known about the general behavior of the Whitney numbers, except for special cases. Stonesifer, for example, has managed to prove that for the lattice of a graphical matroid $(W(2))^2 \geq W(1) W(3)$ [39]. To check logarithmic concavity in the general case, we may apply our general observation made in the section on direct sums: if logarithmic concavity can be proved for connected geometries, it is easy to see that it then holds for their direct sums.

V. TUTTE-GROTHENDIECK INVARIANTS AND THE CRITICAL PROBLEM

An *invariant* is a function on the class of matroids (or geometries) such that $f(G) = f(H)$ if $G \simeq H$. In this section we will consider the wide class of Tutte-Grothendieck invariants. An invariant f is a *Tutte-Grothendieck invariant* if the range of f is a commutative ring, and if $f(G) = f((G - p) \oplus G(p)) = f(G - p)f(p)$ when $p \in G$ is either a loop or an isthmus, and $f(G) = f(G - p) + f(G/p)$ otherwise. We term this method of computing f the *Tutte-Grothendieck recursion*. (An isthmus is a point whose complement is a hyperplane, i.e., the dual of a loop.) It then follows that $f(G \oplus H) = f(G)f(H)$ and that $f(G)$ can be expressed as a unique sum $\sum_{k} f(M_k)$ where each M_k is a minor of G which is a direct sum of isthmuses and loops, $M_k = F^{i,0} \oplus F^{0,j}$, so that $f(M_k) = (f(I))^i (f(L))^j$ where I represents an isthmus and L a loop. Letting $f(I)$ and $f(L)$ be the elements z and x in the ring of polynomials in these two variables, we get a universal invariant, the *Tutte polynomial* $t(G) = f(z, x)$. Every Tutte-Grothendieck (or *T-G*) invariant is then an evaluation of the Tutte polynomial. The commutativity relations at the end of Section II show that if $t(G) = f(z, x)$, then $t(G^*) = f(x, z)$. Other properties of $t(G)$ can be found in [13] and [6]. If we can show that a given invariant h is a Tutte-Grothendieck invariant and if $h(I) = a$ and $h(L) = b$ for an isthmus I and loop L, we obtain $h(G) = f(a, b)$. We then may allow one invariant to give information about another.

Two of the most interesting Tutte-Grothendieck invariants are the subgeometry generating function and (the absolute value of) the characteristic polynomial. For a matroid G on a set S, let a_{ij} be the number of subsets (subgeometries) of G with rank $r(S) - i$ and cardinality $r(S) - i + j$. Then the *subgeometry generating function* $S(G) = \sum_{i,j} a_{ij} u^i v^j$ is a T-G invariant. Further, $S(G) = f(u + 1, v + 1)$. This can be proved most directly by properties of minors and direct sums, noting for example that if A is a subset of G and p is neither an isthmus nor a loop, then the nullity $|A| - r(A)$ and corank $r(S) - r(A)$ of the subgeometry A are the same in G and $G - p$ if $p \notin A$, and are the same for A and $A - p$ in G and G/p respectively if $p \in A$. Evaluations of the subgeometry generating function then yield that $2^{|S|} = f(2, 2)$; the *complexity* or number of bases of G equals $f(1, 1)$ (the number of monomial terms in $t(G)$); the number of independent sets of G equals $f(2, 1)$; and the number of spanning sets (i.e., sets B such that $\overline{B} = S$) equals $f(1, 2)$.

An Ulam-type reconstruction theorem generalized to geometries [8] states that for a matroid G without isthmuses, although we know we cannot in general reconstruct G up to isomorphism from the isomorphism types of its single-point deletions, $\{G - p \mid p \in S\}$, we may reconstruct its subgeometry generating function (and hence its Tutte polynomial) by the formula

$$S(G) = v^{|G|-r(G)} + \sum_{p \in G} \sum_{i,j} \frac{\alpha_{i,j}{}^p}{r(G) + i - j} u^i v^j,$$

where $a_{i,j}{}^p$ is the coefficient of $u^i v^j$ in the subgeometry generating function for $G - p$.

The *characteristic polynomial*, $\chi(G, \lambda)$, computed in the geometric lattice L of G, is a T-G invariante equal to $\pm f(1 - \lambda, 0)$. This is best shown after observing that if $G = G_1 \oplus G_2$, then the geometric lattice of G, $L(G)$, equals the cartesian product $L(G_1) \times L(G_2)$ and in this case $\mu_L(0,(x, y)) = \mu_{L_1}(0, x) \mu_{L_2}(0, y)$; and using the formula (Proposition 5.1 of [35]) that $\mu(0, x) = \sum_{A=x} (-1)^{|A|}$. Details of the above proofs can be found for example in [6]. As a corollary, we obtain that $(-1)^{r(G)} \mu_L(0, 1) = |\mu_L(0, 1)| = f(1, 0) > 0$.

A geometry G of rank n on a set S is a dependence geometry over a finite field F_q if and only if it can be embedded in $P = \mathrm{PG}(n - 1, q)$ as a subgeometry. Although in general this embedding may be accomplished in many ways, an invariant of the embedding is the minimal number c for which there exist hyperplanes $H_1, \ldots, H_c$ in P with $H_1 \cap \cdots \cap H_c \cap S = \emptyset$. This number is the same as the minimal number of linear functionals $f_1, \ldots, f_c$ which distinguish the points of S, i.e., such that for all $p \in S, f_i(p) \neq 0$ for some f_i. This number c is termed [14] the *critical exponent* of G over F_q. The *critical problem* for combinatorial geometries is to determine general methods to compute this critical exponent. We may compute it from $\chi(G, \lambda)$ (and hence from $t(G)$) as follows: $\chi(G, q^i) = (-1)^{r(G)} f(1 - q^i, 0)$ is the number of i-tuples of linear functionals which distinguish the points of S in any embedding of G in $\mathrm{PG}(n - 1, q)$. This can be proved by Möbius inversion [14] or by showing that the number of such i-tuples obeys the T-G recursion [6]. We then have that $\chi(G, q^i)$ is an increasing function of i which is 0 for $i = 1, 2, \ldots, c - 1$, and that $\chi(G, q^c)/(q - 1)^c c!$ is the number of c-subsets of hyperplanes whose intersection with S is empty. Note that $c = 1$ if and only if G can be embedded in $\mathrm{AG}(n - 1, q)$.

Thus if G is a geometry of rank n and critical exponent c, the number $F(k, G)$ of flats of rank $n - k$ in P, whose intersection with G is empty, equals zero for $k < c$, while

$$F(c, G) = \chi(G, q^c) / \prod_{i=0}^{c-1}(q^c - q^i).$$

Recent work [48] has generalized this result to compute from the Tutte polynomial the number I_{rk} of flats of rank r in the ambient projective space which intersect G in k points for all r and k.

As an application of these ideas, the reader may check that if G is the geometry of Example 2, if we first delete the point a (so that e becomes an isthmus), and then contract a, we may compute the Tutte polynomial:

$$t(G) = zt(G_1) + t(G_2) = z^4 + 3z^3 + 5z^2 + z^2x + 2z + 5zx + 2zx^2 + 2x + 2x^2 + x^3$$

where

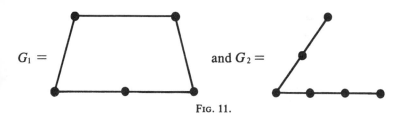

FIG. 11.

Thus G has 128 subsets, 87 independent sets, 24 bases, $\mu(0, 1) = 11$; and over F_3, G spans AG(3, 3) missing precisely four hyperplanes in any embedding into PG(3, 3).

We will now explore some of the applications of the critical problem to the theories of packings and (error-correcting) codes.

The *Hamming metric*, $d(v, w)$, on the vector space F_q^n is defined as the number of nonzero entries in $v - w$. Let $B(n, t)$ be the punctured ball in F_q^n of radius t in the Hamming metric, i.e.,

$$B(n, t) = \{v \in F_q^n \mid 0 < d(0, v) \le t\}.$$

An $(n, n - r)$ *linear code of distance* $t + 1$ is a subspace C of F_q^n of dimension $n - r$ with the property that $C \cap B(n, t) = \emptyset$. The integer r is called the *redundancy* of the code. The *coding problem* is to determine the smallest integer $R(n, t)$ for which there exists an $(n, n - R(n, t))$ code of distance $t + 1$, i.e., to find the smallest feasible redundancy for a code in which the signals have length n and which can "correct" up to $t/2$ transmission errors.

A *t-independent subset* A of the vector space F_q^r has the property that any t elements of A are linearly independent. The *packing problem* is to determine the greatest integer $N(r, t)$ for which there is a t-independent subset of F_q^r of size $N(r, t)$. For example, $N(r, 1) = |F_q^r - \{0\}| = q^r - 1$ and $N(r, 2) = |PG(r, q)| = (q^r - 1)/(q - 1)$.

These two problems are related as follows: Let C be an $(n, n - k)$ code of distance $t + 1$. If M is an $r \times n$ matrix with kernel $k(M) = C$, then $Mv \neq 0$ for all $v \in B(n, t)$; but this latter statement is equivalent to the fact that the columns of M are a t-independent subset of F_q^r. For a fixed t it follows that

$$R(n, t) \leq r \quad \text{iff} \quad N(r, t) \geq n.$$

One can further check that $N(r, t) = \max \{n \mid R(n, t) = r\}$.

These two problems are related to the critical problem in the following manner. Let C be an $(n, n - r)$ code of distance $t + 1$. Let B be a basis for C and extend B by $v_1, \ldots, v_r$ to a basis for F^n. Let H_i be the hyperplane orthogonal to v_i for $i = 1, \ldots, r$. Then $H_1 \cap \cdots \cap H_r \cap B(n, t) = \emptyset$ since any vector in $H_1 \cap \cdots \cap H_r$ must be orthogonal to all of $v_1, \ldots, v_r$ and hence must lie in C. Conversely, for any r hyperplanes with $H_1 \cap \cdots \cap H_r \cap B_{n,t} = \emptyset$ one easily constructs an $(n, n - r)$ code of distance $t + 1$. Therefore $R(n, t)$ is exactly the critical exponent of $B(n, t)$ over F_q. Finally,

$$N(r, t) = \max \{n \mid \text{the critical exponent of } B(n, t) = r\}.$$

This connection between codes, packing, and the critical problem was first noted in [17]; another paper connecting linear codes and matroid theory is [27], where it is shown that the weight enumerator $A_U(z)$ of an $(n, n - r)$ linear code $U \subseteq F_q^n$ can be computed by

$$A_U(z) = (1 - z)^{n-r} z^r f\left(\frac{1 + (q - 1)z}{1 - z}, \frac{1}{z}\right)$$

where $f(z, x) = t(G)$, G being the (column) dependence geometry associated with the rowspace U. This formula and matroid duality verify that MacWilliams duality theorem for codes.

The entire chromatic (and flow) theory of graphs is also an instance of the critical problem. A graph G has a *proper k-coloring* when there is a function $f: V \to K$ such that $|K| = k$ and whenever v and v' are adjacent, $f(v) \neq f(v')$. Given any pair of distinct adjacent vertices v and v', joined by the edge e, proper colorings of $G - e$ in which $f(v) = f(v')$ are in 1-1 correspondence with proper colorings of G/e while proper colorings of $G - e$ in which $f(v) \neq f(v')$ are in 1-1 correspondence with proper colorings of G. Thus if G is one-connected and $e \in G$ is not an isthmus or loop, then G/e and $G - e$ are also one-connected and the number of proper colorings of G, $c(G)$, is equal to $c(G - e) - c(G/e)$, so that

$$(-1)^{r(G)}c(G) = (-1)^{r(G-e)}c(G - e) + (-1)^{r(G/e)}c(G/e).$$

Since the tree (with i isthmuses and j loops) $F^{i,0} \oplus F^{0,j}$ has 0 proper k-colorings if $j > 0$ and $k(k - 1)^i$ otherwise, we obtain the formula: $c(G) = k\chi(G, k) = k(-1)^{r(G)} t(1 - k, 0)$. The four-color theorem can then be restated geometrically: If G is a geometry with forbidden minors L_4, F, $M(K_5)$, $M(K_{3,3})$ and their duals, and if $f(z, x)$ is the Tutte polynomial of G, then $|f(-3, 0)| > 0$.

To illustrate these ideas: a graph is 2^k-colorable if and only if its critical exponent over F_2 is less than or equal to k. For example, a graph G is two-colorable if and only if it may be embedded as a subgeometry of a binary affine space. It is an elementary exercise in linear algebra to show that a binary geometry can be so embedded if and only if it has all even circuits. This provides a geometric proof that G is four-colorable if and only if its associated matroid has critical exponent two over F_2 (if and only if its vertices can be partitioned into two subsets both of whose induced subgraphs have all even circuits).

A further application of the theory to graphs is provided by [38] wherein it is shown that $f(2, 0)$ counts the number of ways to orient the edges of G so that no circuit is oriented cyclically (i.e., so that no closed path can be taken in G which obeys the arrows).

In a related fashion, it can be shown that $f(1, 0)$ (i.e., $\mu(G)$) equals the number of acyclic orientations with a given vertex as the only source. Likewise, $\beta(G)$ (defined as $(\partial f/\partial z(0, 0))$ is the number of acyclic orientations in which a given edge joins the only source to the only sink. This in turn can be used to prove that a geometry G is a connected series-parallel network if and only if $\beta(G) = 1$. The Tutte 5-flow conjecture can be worked out in an analogous fashion: Let G be a directed graph. A (*proper*) k-*flow* of G is a function f from the directed edge set E into nonzero elements of a group with k elements such that for every vertex v

$$\sum_{e \in v^+} f(e) - \sum_{e \in v^-} f(e) = 0,$$

where $e \in v^+$ if e is an edge directed into v and $e \in v^-$ if e is an edge directed away from v. Flows can be seen to be (geometrically) dual to colorings and the number of proper k-flows is equal to $\chi(G^*, k) = |f(0, 1 - k)|$. Tutte has conjectured that every isthmus-free graph has a proper 5-flow, i.e., that $f(0, -4) \neq 0$ for all graphical Tutte polynomials. Jaeger [51] has recently shown, using the theory of matroid union and T-G invariants, that every graph has a proper 8-flow.

Applications exist for other classes of geometries. In [10] it is shown that a transversal geometry has critical exponent at most two in any projective space in which it may be embedded. We mention another application to the theory of affine geometries and convexity.

Let S be a set of points in an affine space $A(n, F)$ over an ordered field F. A subset S' of S can be *separated* if it lies in an open half-space determined by some hyperplane in A with $S - S'$ in the other open half-space. Let p be an extreme point of the convex hull of S. If $S' \subseteq S - p$ can be separated (from $(S - p) - S'$), then either S' or $S' \cup p$ can be separated from its respective S complement. Further, both S' and $S' \cup p$ can be separated if and only if S' can be separated from its $S - p$ complement by a hyperplane passing through p. Thus if $s(S)$ is the number of subsets of S which may be separated, then $s(S) = f(2, 0)$ [46]. Upon consideration of $f(2, 0)$ and $f(2, 2)$, one can show that all subsets of S may be separated if and only if the resulting geometry G is a

Boolean algebra. In fact exactly two subsets cannot be separated if and only if S is a circuit. In particular, some subset of a set with $n + 2$ points in $AG(n, F)$ can always be separated. This is Radon's convexity theorem [29].

These Tutte-Grothendieck methods are generalized to (G, p), a geometry G with a distinguished point p, in [5], leading to a four-variable Tutte polynomial $t((G, p)) = z'f_1(z, x) + x'f_2(z, x)$. This expression has special applications to electrical circuit theory, the edge p distinguishing its two incident vertices which become ports of a network. For example, if G is a graph of unit resistances then $f_1(1, 1)/f_2(1, 1)$ is the resistance in $G - p$ between the two ports. Such an application is not surprising, since for such a network, $G - p'$ can be thought of as giving the edge p' infinite resistance, while contraction changes its resistance to 0.

REFERENCES

1. J. G. Basterfield and L. M. Kelly, "A characterization of sets on n points which determine n hyperplanes", *Proc. Comb. Phil. Soc.*, **64** (1968), 585-588.
2. G. Birkhoff, "Abstract linear dependence in lattices", *Amer. J. Math.*, **57** (1935), 800-804.
3. R. Bixby, "On Reid's characterization of the matroids representable over GF(3)"; P. D. Seymour, "Matroid representation over GF(3)" (to appear, *J. Combinatorial Theory (B)*).
4. R. A. Brualdi, "Comments on bases in dependence structures", *Bull. Austral. Math. Soc.*, **1** (1969), 161-167.
5. T. H. Brylawski, "A combinatorial model for series-parallel networks", *Trans. Amer. Math. Soc.*, **154** (1971), 1-22.
6. ____, "A decomposition for combinatorial geometries", *Trans. Amer. Math. Soc.*, **171** (1972), 235-282.
7. ____, "Some properties of basic families of subsets", *Disc. Math.*, **6** (1973), 333-341.
8. ____, *Reconstructing Combinatorial Geometries*, Springer-Verlag Lecture Notes, vol. 406 (1974), 226-235; "On the nonreconstructibility of combinatorial geometries", *J. Combinatorial Theory (B)* **19** (1975), 72-76.
9. ____, "Modular constructions for combinatorial geometries", *Trans. Amer. Math. Soc.*, **203** (1975), 1-44.
10. ____, "An affine representation for transversal geometries", *Studies in Appl. Math.*, **54** (1975), 143-160.
11. T. H. Brylawski, D. G. Kelly, and T. D. Lucas, *Matroids and Combinatorial Geometries* (preliminary version), Lecture Notes, Department of Mathematics, University of North Carolina, Chapel Hill, N.C. 27514.

12. H. H. Crapo, "Single-element extensions of matroids", *J. Res. Nat. Bur. Standards*, **69B** (1965), 55-65.
13. ___, "The Tutte polynomial", *Aequationes Math.*, **3** (1969), 211-229.
14. H. H. Crapo and G.-C. Rota, *Combinatorial Geometries*, M.I.T. Press, Cambridge, Mass., 1970.
15. ___, "Simplicial geometries", Proc. Symposia in Pure Mathematics, vol. XIX, *Combinatorics*, Amer. Math. Soc., Providence, 1971.
16. P. Crawley and R. P. Dilworth, *Algebraic Theory of Lattices*, Prentice-Hall, Englewood Cliffs, New Jersey, 1973.
17. T. A. Dowling, "Codes, packings, and the critical problem", *Atti del Convegno di Geometria Combinatoria e sue Applicazioni*, Perugia (1972), 209-224.
18. T. A. Dowling and D. G. Kelly, "Elementary strong maps between combinatorial geometries", *Proc. Int. Colloq. in Comb. Th.*, Rome, Italy, 1973, *Atti dei Convegni Lincei*, **17**, Tomo II (1976), 121-152.
19. ___, "Elementary strong maps and transversal geometries", *Disc. Math.*, **7** (1974), 209-224.
20. T. A. Dowling and R. M. Wilson, "The slimmest geometric lattices", *Trans. Amer. Math. Soc.*, **196** (1974), 203-215.
21. ___, "Whitney number inequalities for geometric lattices", *Proc. Amer. Math. Soc.*, **47** (1975), 504-512.
22. R. J. Duffin, "Topology of series-parallel networks", *J. Math. Anal. Appl.*, **10** (1965), 305-318.
23. J. R. Edmonds and D. R. Fulkerson, "Transversal and matroid partition", *J. Res. Nat. Bur. Standards*, **69B** (1965), 147-153.
24. J. Graver, *Lectures on the Theory of Matroids*, University of Alberta, Edmonton, 1966.
25. C. Greene, "A rank inequality for finite geometric lattices", *J. Combinatorial Theory*, **9** (1970), 357-364.
26. ___, "A multiple exchange property for bases", *Proc. Amer. Math. Soc.*, **39** (1973), 45-50.
27. ___, "Weight enumeration and the geometry of linear codes", *Studies in Appl. Math.*, **55** (1976) 119-128.
28. C. Greene and D. Kennedy, "Lectures on combinatorial geometries", *Notes*, Bowdoin College, Brunswick, Maine, 1971.
29. B. Grunbaum and V. Klee, *Convexity and Applications*, CUPM Report No. 16, MAA, 1967.
30. H. Hadwiger, H. Debrunner, and V. Klee, *Combinatorial Geometry in the Plane*, Holt, Rinehart, and Winston, New York, 1964.
31. A. Ingleton and R. Main, "Non-algebraic matroids exist", *Bull. London Math. Soc.*, **7** (1975), 144-146.
32. E. H. Lieb, "Concavity properties and a generating function of Stirling numbers", *J. Combinatorial Theory*, **5** (1968), 203-207.
33. S. Mac Lane, "Some interpretations of abstract linear dependence in terms of projective geometry", *Amer. J. Math.*, **58** (1936), 236-240.
34. R. Rado, "A theorem on independence relations", *Quart. J. Math.*, **13** (1942), 83-89.

35. G.-C. Rota, "On the foundations of combinatorial theory, I", *J. Wahrsch.*, **2** (1966), 340-368.
36. R. Stanley, "Modular elements of geometric lattices", *Alg. Univ.*, **1** (1971), 214-217.
37. ____, "Supersolvable lattices", *Alg. Univ.*, **2** (1972), 197-217.
38. ____, "Acyclic orientation of graphs", *J. Disc. Math.*, **5** (1974), 171-178.
39. J. R. Stonesifer, "Logarithmic concavity for edge lattices of graphs" (to appear).
40. W. T. Tutte, "Lectures on matroids", *J. Res. Nat. Bur. Standards*, **69B** (1965), 1-47.
41. ____, *Introduction to the Theory of Matroids*, American Elsevier, New York, 1970.
42. B. L. van der Waerden, *Moderne Algebra*, Berlin, 1930 and 1937.
43. N. White, "The bracket ring of a combinatorial geometry, I", *Trans. Amer. Math. Soc.*, **202** (1975), 79-95; "The bracket ring of a combinatorial geometry, II", *Trans. Amer. Math. Soc.*, **214** (1975), 1-16.
44. H. Whitney, "2-isomorphic graphs", *Amer. J. Math.*, **55** (1933), 245-254.
45. ____, "On the abstract properties of linear dependence", *Amer. J. Math.*, **57** (1935), 509-533.
46. T. K. Zaslavsky, "Facing up to arrangements: face-count formulas for partitions of space by hyperplanes", *Mem. Amer. Math. Soc.*, **154** (1975).
47. C. Berge, *Graphs and Hypergraphs*, North-Holland, Amsterdam, 1973.
48. T. H. Brylawski, "Intersection theory for embeddings of matroids into uniform geometries" (to appear).
49. T. H. Brylawski and D. Lucas, "Uniquely representable combinatorial geometries", *Atti dei Convegni Lincei*, **17**, *Tomo I* (1976), 83-104.
50. C. Greene and T. L. Magnanti, "Some abstract pivot algorithms", *SIAM J. Appl. Math.* **29** (1975), 530-539.
51. F. Jaeger, "Flows and generalized coloring theorems in graphs" (to appear, *J. Combinatorial Theory (B)*).
52. D. J. A. Welsh, *Matroid Theory*, Academic Press, London, 1976.

COMBINATORIAL CONSTRUCTIONS*

Marshall Hall, Jr.

1. INTRODUCTION

When can we find a subset of triples chosen from a set of v points such that each pair of points occurs in exactly one triple? The triples containing a particular point will together contain every other point exactly once. As each of these contains two other points, the number v of points must be odd. A triple contains three pairs and so the total number of pairs $v(v-1)/2$ must be a multiple of 3. Combining these conditions we see that v must be of one of the forms $v = 6t + 1$ or $v = 6t + 3$. In one of the earliest combinatorial constructions the Reverend T. P. Kirkman [22] in 1847 showed that these necessary conditions were also sufficient and gave a complete construction. By mischance, now hallowed by usage, these are now called Steiner triple systems after Steiner [33] who discussed them first in 1853.

---

*This research was supported in part by NSF Grant GP 36230X.

Steiner triple systems are one instance of a general incidence structure, consisting of objects P, usually called points, and blocks B together with an incidence relation PIB between some of them. In section 2 these and some other combinatorial systems including Latin squares and Hadamard matrices are defined and illustrated.

Groups, graphs, geometries, and certain number theoretical constructs all have a combinatorial side, and in section 3 some indication is given of the role of these other subjects in direct constructions of combinatorial systems.

In section 4 an account is given of the methods of constructing recursively large combinatorial systems from smaller ones.

These methods have resolved two major conjectures:

EULER CONJECTURE (1782): *There does not exist a pair of Graeco-Latin squares of order n when n is of the form $n = 4t + 2$.*

This conjecture was completely demolished in 1959 by Bose, Parker, and Shrikhande [11, 12, 27] showing that apart from $n = 2$ and $n = 6$ the conjecture is false.

EXISTENCE CONJECTURE: *Given the block size k and index λ the necessary conditions $\lambda(v - 1) \equiv 0 \pmod{k - 1}$ and $\lambda v(v - 1) \equiv 0 \pmod{k(k - 1)}$ for the existence of a partially balanced incomplete block design are sufficient if v is sufficiently large.*

This, and indeed a more general result, was proved by Richard Wilson [39] in 1973.

2. COMBINATORIAL SYSTEMS. DEFINITIONS AND EXAMPLES

Many combinatorial systems are arrangements of elements p_i, called *points* into sets B_j called *blocks* or *lines*, but blocks are distinguished by their indices, so that different blocks may indeed contain the same points. Such a system is called an *incidence structure* which we now define precisely.

An incidence structure S is a triple $(\{p\}, \{B\}, I)$ where $\{p\}$, $\{B\}$, and I are sets with

$$\{p\} \cap \{B\} = \emptyset, \text{ and } I \subseteq \{p\} \times \{B\}. \tag{2.1}$$

The elements $\{p\}$ are called points, those of $\{B\}$ are called blocks, and we say that p_i is incident with B_j if (p_i, B_j) is one of the pairs of I. We also write this as $p_i \, I \, B_j$. It is also convenient to describe incidence in other ways, saying that p_i is on B_j, or that B_j contains p_i.

An incidence structure is so general that almost always we consider systems with further properties. Dembowski [10] and others call an incidence structure a *tactical configuration* if each of b blocks contains the same number, k, of points and if each of v points lies on the same number, r, of blocks. Counting the total number of incidences by blocks and also by points we have

$$bk = vr. \tag{2.2}$$

An example of a tactical configuration with $b = 8$, $k = 4$, $v = 16$, $r = 2$ is the following, where the points are $\{1, 2, \ldots, 16\}$ and the blocks are B_j, $j = 1, \ldots, 8$. For each block we list the points on it.

$$\begin{aligned}
&B_1: 1, 2, 3, 4 \\
&B_2: 1, 5, 6, 7 \\
&B_3: 2, 8, 9, 10 \\
&B_4: 3, 11, 12, 13 \\
&B_5: 4, 14, 15, 16 \\
&B_6: 5, 8, 11, 14 \\
&B_7: 6, 9, 12, 15 \\
&B_8: 7, 10, 13, 16.
\end{aligned} \tag{2.3}$$

A tactical configuration in which every pair of distinct points occurs together exactly λ times is called a *balanced incomplete block design*. These have been widely used by statisticians in the design of experiments, and an excellent account of this may be found in the book by Mann [25]. Because of their importance, balanced incomplete block designs will be referred to simply as block designs, or designs.

A block design D with b blocks each containing k points, and each of v points lying on r blocks, and with every pair of distinct points on λ blocks will have its parameters v, b, r, k, λ satisfying the following conditions:

$$bk = vr, \; r(k-1) = \lambda(v-1). \tag{2.4}$$

Unless $k = v$ and every block contains all the points we will also have

$$b \geq v. \tag{2.5}$$

This is known as Fisher's inequality after the statistician who first discovered it in 1940 [11]. Associated with a design D is its incidence matrix $A = [a_{ij}]\; i = 1, \ldots, v,\; j = 1, \ldots, b$, in which, the points being $p_1, \ldots, p_v$, and blocks being $B_1, \ldots, B_b$, we have $a_{ij} = 1$ if $p_i \; I \; B_j$ and $a_{ij} = 0$ otherwise. Here A^T being the transpose of A, we have

$$AA^T = (r - \lambda)I + \lambda J, \tag{2.6}$$

where I is the v by v identity matrix and J is the v by v matrix all of whose entries are 1's. Here it is easily shown that the determinant of the matrix on the right hand side of (2.6) is $(r - \lambda)^{v-1}((v-1)\lambda + r)$ and so is non-singular if $r > \lambda$. Here in any event $r \geq \lambda$ since a point p_i cannot appear with another point p_s λ times unless p_i occurs at least λ times in the blocks. But if $r = \lambda$ then every time p_i appears in a block it is paired with every other point and so we have the case $v = k$ which is somewhat trivial. Since the rank of AA^T which is v, is at most b, the Fisher inequality $b \geq v$ follows. This and other properties of designs associated with the incidence matrix A are treated at some length in the article "Combinatorial Matrix Theory" by H. J. Ryser in this volume and in his book [31].

A design for which $b = v$ and so also $r = k$ is called a *symmetric block design*. Since symmetric designs include finite projective planes Dembowski has used the term projective for these designs, but this terminology is not widely used.

An example of a block design is the symmetric design with $b = v = 7$, $r = k = 3$, and $\lambda = 1$ is the following:

$$
\begin{aligned}
&B_0: 1, 2, 4 \\
&B_1: 2, 3, 5 \\
&B_2: 3, 4, 6 \\
&B_3: 4, 5, 0 \\
&B_4: 5, 6, 1 \\
&B_5: 6, 0, 2 \\
&B_6: 0, 1, 3.
\end{aligned}
\tag{2.7}
$$

Another symmetric design with $b = v = 11$, $r = k = 5$, $\lambda = 2$ is the following:

$$
\begin{aligned}
&B_0: 1, 3, 4, 5, 9 & &B_6: 7, 9, 10, 0, 4 \\
&B_1: 2, 4, 5, 6, 10 & &B_7: 8, 10, 0, 1, 5 \\
&B_2: 3, 5, 6, 7, 0 & &B_8: 9, 0, 1, 2, 6 \\
&B_3: 4, 6, 7, 8, 1 & &B_9: 10, 1, 2, 3, 7 \\
&B_4: 5, 7, 8, 9, 2 & &B_{10}: 0, 2, 3, 4, 8. \\
&B_5: 6, 8, 9, 10, 3 & &
\end{aligned}
\tag{2.8}
$$

A design which is not symmetric is the following with $v = 9$, $b = 12$, $r = 4$, $k = 3$, $\lambda = 1$:

$$
\begin{array}{lllll}
B_1: 1, 2, 3 & B_4: 1, 4, 7 & B_7: 1, 5, 8 & B_{10}: 1, 6, 9 \\
B_2: 4, 5, 6 & B_5: 2, 5, 9 & B_8: 2, 6, 7 & B_{11}: 2, 4, 8 \\
B_3: 7, 8, 9 & B_6: 3, 6, 8 & B_9: 3, 4, 9 & B_{12}: 3, 5, 7.
\end{array}
\tag{2.9}
$$

Recently a generalization of block designs has been introduced called a *partially balanced design* abbreviated PBD. A partially balanced design $D(v, K, \lambda)$, $K = \{k_1, k_2, \ldots,\}$ is an incidence structure with v points whose block sizes are chosen from a set $K = \{k_1, k_2, \ldots,\}$ with the property that any pair of distinct points occurs together in exactly λ blocks. From a partially balanced design on v points we may construct another on $v - t$ points by deleting t points. The value of λ will not be changed but we may have further block sizes. Thus deleting one point from the design of (2.8) gives us a design $D(10, K, 2)$ with $K = \{4, 5\}$, while deleting 2 points gives a design $D(9, K, 2)$ with $K = \{3, 4, 5\}$ while if we delete all 5 points of a block and the block itself we have a $D(6, K, 2)$ where $K = \{3\}$ and this is a block design with $v = 6$, $b = 10$, $r = 5$, $k = 3$, $\lambda = 2$.

A Latin square L of order n is an n by n square whose entries are n different elements (usually the numbers $1, 2, \ldots, n$) such that every element occurs exactly once in each row and exactly once in each column. Two n by n Latin squares L_1 and L_2 are said to be *orthogonal* if when we take for the cell in row i and column j the pair $(r, s)_{ij}$ where r is the entry in this cell in L_1 and s is the entry in this cell in L_2, then the n^2 pairs $(r, s)_{ij}$ as $i, j = 1, \ldots, n$ make up all possible n^2 pairs (u, v) of elements. Latin squares $L_1, L_2, \ldots, L_m$ are said to be mutually orthogonal if any two of them are orthogonal.

Here are three mutually orthogonal 4 by 4 squares:

$$
\begin{array}{cccc}
L_1 & & L_2 & & L_3
\end{array}
$$

$$
\begin{array}{cccc cccc cccc}
1 & 2 & 3 & 4 & 1 & 2 & 3 & 4 & 1 & 2 & 3 & 4 \\
2 & 1 & 4 & 3 & 3 & 4 & 1 & 2 & 4 & 3 & 2 & 1 \\
3 & 4 & 1 & 2 & 4 & 3 & 2 & 1 & 2 & 1 & 4 & 3 \\
4 & 3 & 2 & 1 & 2 & 1 & 4 & 3 & 3 & 4 & 1 & 2.
\end{array}
\quad (2.10)
$$

Two rows $(a_1, \ldots, a_j, \ldots, a_{n^2})$ and $b_1, \ldots, b_j, \ldots, b_{n^2})$ where the a's and b's are the numbers $1, 2, \ldots, n$ are said to be *orthogonal* if the pairs $(a_j, b_j), j = 1, \ldots, n^2$ together make up all n^2 pairs (u, v), $u, v = 1, \ldots, n$. An *orthogonal array* $OA(n, t)$ is a rectangle $A = [a_{ij}]$ with t rows and n^2 columns in which any two rows are orthogonal. Here is an $OA(4, 5)$:

$$\begin{array}{lllllllllllllllll}
R_1 & 1&1&1&1 & 2&2&2&2 & 3&3&3&3 & 4&4&4&4 \\
R_2 & 1&2&3&4 & 1&2&3&4 & 1&2&3&4 & 1&2&3&4 \\
R_3 & 1&2&3&4 & 2&1&4&3 & 3&4&1&2 & 4&3&2&1 \\
R_4 & 1&2&3&4 & 3&4&1&2 & 4&3&2&1 & 2&1&4&3 \\
R_5 & 1&2&3&4 & 4&3&2&1 & 2&1&4&3 & 3&4&1&2.
\end{array}$$
(2.11)

A set of h mutually orthogonal n by n squares L_1, L_2, ..., L_h yields an OA(n, $h + 2$) in the following way: In column $C_{(n-1)i+j} = C_{i,j}$ let the entry in row R_1 be i and in row R_2 be j. Construct row R_{m+2} from L_m putting x in column $C_{i,j}$ if in L_m the entry $L_m(i, j) = x$ this being the entry in the ith row and jth column of L_m. Thus in the orthogonal array A we put $a_{(m-1)i+j,m+2} = L_m(i, j)$. Here R_1 and R_2 are orthogonal as constructed. The orthogonality of R_{2+m} and R_1 (alternately R_2) says that each digit 1, 2, ..., n occurs exactly once in each row (alternately column). The orthogonality of R_{2+m} and R_{2+k} is equivalent to the orthogonality of squares L_m and L_k. The OA(4, 5) of (2.11) has been constructed in this way from the Latin squares L_1, L_2, L_3 of (2.10). Conversely from an OA(m, $h + 2$) we may readily construct h mutually orthogonal Latin squares L_1, ..., L_k by first permuting the columns so that column $C_{(n-1)i+j}$ has i in row R_1 and j in row R_2 and then defining L_m from R_{m+2} by putting $L_m(i, j) = a_{(n-1)i+j,m+2}$. The properties of the OA(n, $h + 2$) then assure us that L_1, ..., L_h are mutually orthogonal squares.

A construct essentially the same as an orthogonal array is *transversal design* TD(k, n) with k *groups* of size n. Here is meant a triple (X, $\mathcal{G}$, $\mathcal{C}$) where X is a set of kn points, $\mathcal{G} = \{G_1, ..., G_k\}$ is a partition of X into k subsets G_i (called *groups*) each containing n points, and $\mathcal{C}$ is a class of subsets of X (called *blocks* or *transversals*) such that each block $A \in \mathcal{C}$ contains precisely one point from each group and each pair of points not contained in the same group occur together in precisely one block A. Clearly each block contains k points and it is not difficult to see that each point occurs in n blocks and the total number of blocks is n^2. We may construct a TD(k, n) from an OA(n, k) in the following manner: The points of X are x_{rs}, $r = 1, ..., k$, $s = 1, ..., n$. A row R_i of OA(n, k) determines the group G_i of TD(k, n) by taking G_i as the points $x_{i1}, ..., x_{in}$ corresponding to the digits 1, 2, ..., n of R_i. A block (or transversal) A_j is determined by a column C_j of OA(n, k)

where if $a_{ij} = d$ in OA(n, k) we assign the point x_{id} to the transversal A_j. The properties of the TD(k, n) are now immediate from those of the OA(n, k) and we may clearly reverse the construction. Corresponding to the OA(4, 5) of (2.11) we have the following TD(4, 5) where we use different letters for the points rather than double subscripts:

$$\text{Groups:} \quad G_1: a_1, a_2, a_3, a_4$$
$$G_2: b_1, b_2, b_3, b_4$$
$$G_3: c_1, c_2, c_3, c_4$$
$$G_4: d_1, d_2, d_3, d_4$$
$$G_5: e_1, e_2, e_3, e_4$$

Transversals: (2.12)

A_1: $a_1\ b_1\ c_1\ d_1\ e_1$ A_9: $a_3\ b_1\ c_3\ d_4\ e_2$

A_2: $a_1\ b_2\ c_2\ d_2\ e_2$ A_{10}: $a_3\ b_2\ c_4\ d_3\ e_1$

A_3: $a_1\ b_3\ c_3\ d_3\ e_3$ A_{11}: $a_3\ b_3\ c_1\ d_2\ e_4$

A_4: $a_1\ b_4\ c_4\ d_4\ e_4$ A_{12}: $a_3\ b_4\ c_2\ d_1\ e_3$

A_5: $a_2\ b_1\ c_2\ d_3\ e_4$ A_{13}: $a_4\ b_1\ c_4\ d_2\ e_3$

A_6: $a_2\ b_2\ c_1\ d_4\ e_3$ A_{14}: $a_4\ b_2\ c_3\ d_1\ e_4$

A_7: $a_2\ b_3\ c_4\ d_1\ e_2$ A_{15}: $a_4\ b_3\ c_2\ d_4\ e_1$

A_8: $a_2\ b_4\ c_3\ d_2\ e_1$ A_{16}: $a_4\ b_4\ c_1\ d_3\ e_2$

In (2.12) if we consider both the groups and the transversals as blocks we have a pairwise balanced design $D(20, K, 1)$ where $K = \{4, 5\}$.

A geometrical interpretation may be given to a set of mutually orthogonal squares. Let us identify a point (i, j) with the cell in the ith row and jth column of each square, $i, j = 1, \ldots, n$. For points (x, y) take as lines the n points with $x = i$, and as $i = 1, \ldots, n$ giving a family of n parallel lines. Similarly $y = j$, $j = 1, \ldots, n$ gives a second family of parallel lines. A Latin square L_m now defines a further family of parallel lines $M_1, \ldots, M_n$, those points (x, y) lying on the line M_s for which the entry $L_m(x, y) = s$. Orthogonality of the squares is equivalent to the requirement that two distinct lines cannot contain two distinct points. Thus h mutually orthogonal n by n squares or an OA($n, h + 2$) correspond to $h + 2$ families of n parallel lines each with n points.

If $q = p^r$ is any prime power then the affine plane over GF(q) yields $q + 1$ families of parallel lines and so $q + 1$ mutually orthogonal Latin squares of order q. The q^2 points are (x, y) $x, y \in$ GF(q). One family of parallels consists of the lines $x = c$, $c \in$ GF(q) Other lines contain the points (x, y) satisfying an equation $y = mx + b$. For each fixed "slope" m the corresponding lines form a family of parallel lines.

For designs with $\lambda = 1$ the *group divisible design* GDD($v, \mathcal{G}, K, 1$) is a generalization of the transversal design. Here the groups $G_1, G_2, \ldots, G_m$ are a partition of the v points, not necessarily of the same size, and other blocks have at most one point in a group G_i. Considering the groups as blocks we may add a new point to all of the groups and obtain a pairwise balanced design with $\lambda = 1$. Sometimes a block design can be considered a group divisible design in more than one way using different blocks as the groups each time. In this case we may add a new point for each set of groups adding it to all groups in the set. Then we may also add a further block containing all the new points and obtain a pairwise balanced design. A strong form of this situation arises in what is called a resolvable design.

A design $D(v, K, \lambda)$ is said to be *resolvable* if the blocks can be partitioned into sets $R_1, \ldots, R_s$ such that the blocks in a set R_i are a partition of the points of D. If $\lambda = 1$ we can add a point P_i to each of the blocks in the set R_i, $i = 1, \ldots, t$ and also adjoin a new block whose points are $P_1, \ldots, P_t$ to obtain a pairwise balance design on $v + t$ points. The design of (2.9) with $v = 9$, $b = 12$, $r = 4$, $k = 3$, $\lambda = 1$ is resolvable, there being 4 sets of 3 blocks each set containing all 9 points. Adding one point we have a pairwise balanced design $D(10, K, 1)$, $K = \{4, 3\}$; adding two, three, or four points and a block through them we have respectively a $D(11, K, 1)$, $K = \{4, 3, 2\}$, $D(12, K, 1)$, $K = \{4, 3\}$ and a $D(13, K, 1)$ with $K = \{4\}$.

An Hadamard matrix $H = [h_{ij}]$ is a square matrix of order n with $h_{ij} = \pm 1$ which satisfies the matrix equation

$$HH^T = H^TH = nI_n. \qquad (2.13)$$

Here H^T is the transpose of H and I_n is the identity matrix of order n:
The matrices

$$[1], \begin{bmatrix} 1 & 1 \\ 1 & - \end{bmatrix}, \begin{bmatrix} 1 & 1 & 1 & 1 \\ 1 & - & 1 & - \\ 1 & 1 & - & - \\ 1 & - & - & 1 \end{bmatrix}, \begin{bmatrix} - & 1 & 1 & 1 \\ 1 & - & 1 & 1 \\ 1 & 1 & - & 1 \\ 1 & 1 & 1 & - \end{bmatrix}, \qquad (2.14)$$

are Hadamard matrices of orders 1, 2, and 4 respectively; where we write $-$ for -1.

Clearly we may permute rows or columns or change the sign of rows or columns of an Hadamard matrix and obtain another Hadamard matrix. As 2.13 says that any two rows (or columns) of an Hadamard matrix have inner product zero, let us consider the pattern of the first three rows, changing the signs of columns to make the first row all $+1$'s.

$$\begin{array}{c|c|c|c}
r & s & t & u \\
\hline
1 \cdots 1 & 1 \cdots 1 & 1 \cdots 1 & 1 \cdots 1 \\
1 \cdots 1 & 1 \cdots 1 & - \cdots - & - \cdots - \\
1 \cdots 1 & - \cdots - & 1 \cdots 1 & - \cdots -
\end{array} \qquad (2.15)$$

Here r is the number of columns of form $\begin{bmatrix} 1 \\ 1 \\ 1 \end{bmatrix}$ and s, t, u are similarly defined. Then we have

$$\begin{aligned} r + s + t + u &= n \\ r + s - t - u &= 0 \\ r - s + t - u &= 0 \\ r - s - t + u &= 0 \end{aligned} \qquad (2.16)$$

The first of these says that the total number of columns is n, the second, third, and fourth that the inner products respectively of the first and second, first and third, and second and third rows is zero. From these it is immediate that $r = s = t = u = n/4$, and so an Hadamard matrix of order at least 3 must have order a multiple of 4. It is conjectured that there is an Hadamard matrix of every order n which is a multiple of 4. The smallest order in doubt as this is written is $n = 268$.

We may normalize an Hadamard matrix H by changing the signs of rows and columns so that the first row and first column consist entirely of $+1$'s. If H is of order $n = 4t$, by eliminating the first row and first column the remaining $4t - 1$ rows each contains $2t - 1$ $+1$'s. This matrix $A = [a_{ij}]$ determines $4t - 1$ blocks B_i each containing $2t - 1$ points namely the set of j's for which $a_{ij} = +1$. These blocks form a symmetric design D_H with parameters

$$v = 4t - 1, k = 2t - 1, \lambda = t - 1. \qquad (2.17)$$

The fact that any pair of points occurs together in $\lambda = t - 1$ blocks is an immediate consequence of the orthogonality of the columns of H. Conversely given a symmetric design with the parameters of (2.17) we may construct the matrix $A = [a_{ij}]$ putting $a_{ij} = +1$ if the jth point is in the ith block and $a_{ij} = -1$ otherwise. Then bordering the matrix A with a first row and column of $+1$'s the resulting matrix is an Hadamard matrix H. The design (2.8) is of the type (2.17) and the corresponding Hadamard matrix H_{12} is given by

COMBINATORIAL CONSTRUCTIONS   229

$$H_{12} = \begin{array}{c|cccccccccccc} & \infty & 0 & 1 & 2 & 3 & 4 & 5 & 6 & 7 & 8 & 9 & 10 \\ \hline \infty & 1 & 1 & 1 & 1 & 1 & 1 & 1 & 1 & 1 & 1 & 1 & 1 \\ 0 & 1 & - & 1 & - & 1 & 1 & 1 & - & - & - & 1 & - \\ 1 & 1 & - & - & 1 & - & 1 & 1 & 1 & - & - & - & 1 \\ 2 & 1 & 1 & - & - & 1 & - & 1 & 1 & 1 & - & - & - \\ 3 & 1 & - & 1 & - & - & 1 & - & 1 & 1 & 1 & - & - \\ 4 & 1 & - & - & 1 & - & - & 1 & - & 1 & 1 & 1 & - \\ 5 & 1 & - & - & - & 1 & - & - & 1 & - & 1 & 1 & 1 \\ 6 & 1 & 1 & - & - & - & 1 & - & - & 1 & - & 1 & 1 \\ 7 & 1 & 1 & 1 & - & - & - & 1 & - & - & 1 & - & 1 \\ 8 & 1 & 1 & 1 & 1 & - & - & - & 1 & - & - & 1 & - \\ 9 & 1 & - & 1 & 1 & 1 & - & - & - & 1 & - & - & 1 \\ 10 & 1 & 1 & - & 1 & 1 & 1 & - & - & - & 1 & - & - \end{array}$$

(2.18)

If H and K are Hadamard matrices of orders n and m respectively then an Hadamard matrix of order nm is easily constructed by taking the Kronecker product of H by K, $H \times K$ (this is sometimes called the direct product or the tensor product). This is the block matrix

$$H \times K = \begin{bmatrix} h_{11}K, & h_{12}K, & \ldots, & h_{1n}K \\ h_{21}K, & h_{22}K, & \ldots, & h_{2n}K \\ \vdots & \vdots & & \vdots \\ h_{n1}K, & h_{n2}K, & \ldots, & h_{nn}K \end{bmatrix} \quad (2.19)$$

Taking $H = K = \begin{bmatrix} 1 & 1 \\ 1 & - \end{bmatrix}$ we obtain the third matrix of (2.14) in this way.

Another type of symmetric design can be associated with an Hadamard matrix, necessarily of square order, without bordering. If we have

$$v = 4t^2, \quad k = 2t^2 - t, \quad \lambda = t^2 - t. \tag{2.20}$$

Here take the $4t^2$ blocks as rows and define $H = [h_{ij}]$ putting $h_{ij} = +1$ if the jth point is in the ith block, and $h_{ij} = -1$ otherwise for $i, j = 1, \ldots, 4t^2$.

If we take $H = -K = \begin{bmatrix} - & 1 & 1 & 1 \\ 1 & - & 1 & 1 \\ 1 & 1 & - & 1 \\ 1 & 1 & 1 & - \end{bmatrix}$ than $H \times K$ is an

Hadamard matrix H_{16} of order 16 corresponding in this way to a symmetric design with $v = 16$, $k = 6$, $\lambda = 2$.

Also since there is an Hadamard matrix of order 2, whenever there exists an H_n there also exists an H_{2n} so that for $n = 4m$ we need only concern ourselves with cases for which m is odd, in considering the general question of existence.

3. DIRECT CONSTRUCTIONS OF COMBINATORIAL SYSTEMS

Combinatorial systems can be constructed in a number of ways from finite fields, finite geometries, and finite groups. There are also some constructions based on number theoretical constructs and some algorithms for construction. These methods we call direct constructions. Other methods involve compositions of systems to form larger systems, and will be called recursive constructions. These will be discussed in the next section.

Two incidence structures S_1 and S_2 are said to be *isomorphic* if there is one-to-one mapping α, called an isomorphism, of points of S_1 onto those of S_2 and of the blocks of S_1 onto those of S_2 such that incidence is preserved, i.e., that if and only if (p, B) is an incidence in S_1, then $(p\alpha, B\alpha)$ is an incidence in S_2. An isomorphism of an incidence structure with itself is called an *automorphism*. Clearly the automorphisms of an incidence structure form a group.

The symmetric designs (2.7) and (2.8) both have automorphism groups which permute the points and the blocks in a single cycle of length $v = b$. For (2.7) we have $\alpha = (0, 1, 2, 3, 4, 5, 6) (B_0, B_1, B_2, B_3, B_4, B_5, B_6,)$ and for (2.8) we have $\alpha = (0, 1, \ldots, 10) (B_0, B_1, \ldots, B_{10})$, where we have written the automorphism in cycle form in both cases. A symmetric v, k, λ design D is called a cyclic design if there is an automorphism α moving the v points and the v blocks in a cycle of length v. The block B_5: 6, 8, 9, 10, 3 of (2.8) has the property that the $k(k-1) = 5 \cdot 4 = 20$ non-zero differences of its elements modulo v, $v = 11$ yield every $d \not\equiv 0 \pmod{11}$ exactly λ times, $\lambda = 2$. For

$$\begin{aligned}
1 &\equiv 9 - 8 \equiv 10 - 9, & 6 &\equiv 3 - 8 \equiv 9 - 3, \\
2 &\equiv 8 - 6 \equiv 10 - 8, & 7 &\equiv 6 - 10 \equiv 10 - 3, \\
3 &\equiv 9 - 6 \equiv 6 - 3, & 8 &\equiv 6 - 9 \equiv 3 - 6, \quad (3.1) \\
4 &\equiv 10 - 6 \equiv 3 - 10, & 9 &\equiv 6 - 8 \equiv 8 - 10, \\
5 &\equiv 8 - 3 \equiv 3 - 9, & 10 &\equiv 8 - 9 \equiv 9 - 10.
\end{aligned}$$

DEFINITION: *A set of k residues $\{a_1, a_2, \ldots, a_k\}$ modulo v is called a (v, k, λ) difference set if for every $d \not\equiv 0 \pmod{v}$ there are exactly λ ordered pairs a_i, a_j such that $a_i - a_j \equiv d \pmod{v}$.*

Thus

$$6, 8, 9, 10, 3 \qquad (3.2)$$

are an $(11, 5, 2)$ difference set. Note that from the definition of a difference set we must have $k(k-1) = \lambda(v-1)$.

THEOREM 3.1: *A set of k residues $\{a_1, \ldots, a_k\}$ modulo v is a (v, k, λ) difference set if and only if the sets B_i: $\{a_1 + i, a_2 + i, \ldots, a_k + i\}$ modulo v, $i = 0, \ldots, v - 1$, are a cyclic (v, k, λ) block design D.*

Proof: Suppose first that the sets B_i: $\{a_1 + i, a_2 + i, \ldots, a_k + i\}$ form a (v, k, λ) design D. Then for $d \not\equiv 0 \pmod{v}$ the pair of points $d, 0$ occur together exactly λ times, so that there are λ blocks B_t containing both d and 0, so that for appropriate a_i, a_j we have $a_i + t \equiv d$, $a_j + t \equiv 0$ and so $a_i - a_j \equiv d \pmod{v}$ while if $a_i - a_j \equiv d \pmod{v}$ we can determine t by $a_j + t = 0$ and so $a_i + t = d$. Hence if the sets B_i form a design D then $\{a_1, \ldots, a_k\}$ is a (v, k, λ) difference set. Clearly D is a cyclic design with automorphism α: $i \to i + 1 \pmod{v}$ for points and $B_i \to B_{i+1}$ for blocks. Conversely suppose that $\{a_1, \ldots, a_k\}$ is a (v, k, λ) difference set. Then define the sets B_i: $\{a_1 + i, \ldots, a_k + i\}$ modulo v, $i = 0, \ldots, v - 1$. If r, s are distinct residues modulo v, let $r - s \equiv d \pmod{v}$ and let $a_i - a_j \equiv d \pmod{v}$. Determine t by $a_j + t \equiv s \pmod{v}$. Then clearly $a_i + t \equiv d + a_j + t \equiv s + d \equiv r \pmod{v}$ and both r and s are in the set B_t. Since there are λ choices of a_i and a_j with $a_i - a_j \equiv d \pmod{v}$, this shows that every pair of distinct points occurs together in λ sets B_t.

More generally, if G is an Abelian group of order v written additively and if $\{a_1, \ldots, a_k\}$ is a difference set of k distinct elements of G such that for every $d \neq 0$ of G there are exactly λ solutions to $a_i - a_j = d,$, then the sets B_u $\{a_1 + u, a_2 + u, \ldots, a_k + u\}$ $u \in G$ form a (v, k, λ) block design D for which G is an automorphism group in which no element of G except the zero element fixes any point or block. Here we call D an Abelian design. It is not difficult to extend the concept of difference set to a non-Abelian group [2] but up to the present the results have not been particularly satisfying.

If G is an Abelian group of order v and if t is an integer relatively prime to v, then the mapping $x \to tx$, $x \in G$ is an automorphism of G. It may happen that this automorphism is an automorphism of an Abelian design. If this is the case we call t a multiplier of the difference set.

DEFINITION: *If D is an Abelian (v, k, λ) design with points the elements of the Abelian group G of order v, and if t is an integer with (t, v) = 1 such that $x \to tx$, $x \in G$ is an automorphism of D, then t is called a multiplier of D.*

It is a remarkable fact that every known cyclic design has a non-trivial multiplier. More generally we may follow Bruck and call an automorphism of G which is also an automorphism of an Abelian design over G a multiplier. The existence of a multiplier is of great assistance in constructing an Abelian design or showing that it does not exist. Note that 2 is a multiplier of (2.7) and 3 is a multiplier of (2.8).

If $\{a_1, \ldots, a_k\}$ is a (v, k, λ) difference set over an Abelian group G of order v, then it is easily checked that t is a multiplier if and only if there is an $s \in G$ such that the following sets are equal in some order:

$$\{ta_1, ta_2, \ldots, ta_k\} = \{a_1 + s, a_2 + s, \ldots, a_k + s\}. \qquad (3.3)$$

The following theorem due to Hall and Ryser [16] proves the existence of multipliers in many cases:

THEOREM 3.2 (Multiplier theorem): *If $\{a_1, \ldots, a_k\}$ are a (v, k, λ) difference set over the Abelian group G of order v and if p is a prime such that (i) p divides $n = k - \lambda$ (ii) $(p, v) = 1$ and (iii) $p > \lambda$, then p is a multiplier of the Abelian design D.*

Proof: We write the group G in multiplicative form and consider the group ring $R = R(G)$ over the rational integers Z. Thus if $x_1 = 1, x_2, \ldots, x_v$ are the elements of G the elements of R are $b_1 x_1 + \cdots + b_v x_v$ with the b's rational integers. In R the sum of elements is obtained by adding coefficients, the products in R being determined by $x_i x_j = x_s$ in G and the distributive laws. If G is the additive group of residues $0, 1, \ldots, v - 1 \pmod{v}$, the multiplicative form of G is $1 = x^0, x, x^2, \ldots, x^{v-1}$ with $x^v = 1$.

If $\{a_1, \ldots, a_k\}$ is our difference set in additive form, let $d_1, \ldots, d_k$ be the corresponding group elements in multiplicative form and let us define $\theta(d)$ by

$$\theta(d) = d_1 + d_2 + \cdots + d_k. \tag{3.4}$$

Note that the addition here is the addition of the group ring R and not the group operation of G. Let us define an element T of R by

$$T = \sum_{x \in G} x = x_1 + x_2 + \cdots + x_v. \tag{3.5}$$

For any exponent t let us define

$$\theta(d^t) = d_1^t + d_2^t + \cdots + d_k^t. \tag{3.6}$$

Then the following relation expresses the fact that $\{a_1, \ldots, a_k\}$ is a difference set:

$$\theta(d)\theta(d^{-1}) = (k - \lambda) \cdot 1 + \lambda T = n + \lambda T \tag{3.7}$$

writing n for $k - \lambda$.

For $\theta(d)\theta(d^{-1}) = \sum_{i,j=1}^{k} d_i d_j^{-1}$. Here there are k terms $d_i d_j^{-1}$ equal to the identity 1 and as $d_i d_j^{-1}$ is the multiplicative form of $a_i - a_j$, every non identity element x of G is expressible in the form $d_i d_j^{-1}$ in exactly λ ways. Hence with $x_1 = 1$, $\theta(d)\theta(d^{-1}) = k \cdot x_1 + \lambda x_2 + \cdots + \lambda x_v = (k - \lambda)1 + \lambda T$ as given in (3.7).

Since G is Abelian and the binomial coefficients $\binom{p}{i}$, $i = 1, 2, \ldots, p-1$, are multiples of p we have found for elements of R

$$(A + B)^p = A^p + B^p + pW, \tag{3.8}$$

where W is some element of R. Applying this repeatedly we have

$$\theta(d)^p = d_1^p + d_2^p + \cdots + d_p^p + pM = \theta(d^p) + pM. \tag{3.9}$$

Multiplying (3.7) by $\theta(d)^{p-1}$ we have

$$\theta(d)^p \theta(d^{-1}) = n\theta(d)^{p-1} + \lambda \theta(d)^{p-1} T. \tag{3.10}$$

For $x \in G$ we have $xT = T$ and so $(b_1x_1 + b_2x_2 + \cdots + b_vx_v)T = (b_1 + b_2 + \cdots + b_v)T$. Hence $\theta(d)T = kT$ and $\theta(d)^{p-1}T = k^{p-1}T$. Thus (3.10) becomes

$$\theta(d)^p\theta(d^{-1}) = n\theta(d)^{p-1} + \lambda(k^{p-1} - 1)T + \lambda T. \tag{3.11}$$

Now p is a divisor of $n = k - \lambda$. If p divides k, then p also divides λ. If p does not divide k, then p divides $k^{p-1} - 1$. In either event p divides $\lambda(k^{p-1} - 1)$ and so we write (3.1) in the form

$$\theta(d)^p\theta(d^{-1}) = pH + \lambda T. \tag{3.12}$$

From (3.9) we have

$$(\theta(d^p) + pM)\theta(d^{-1}) = pH + \lambda T, \tag{3.13}$$

which gives

$$\theta(d^p)\theta(d^{-1}) = pS(x) + \lambda T \tag{3.14}$$

for $S(x)$ some element of R.

If $S(x) = s_1x_1 + \cdots + s_vx_v$, the homomorphism of G onto the identity $x \to 1$ maps $\theta(d^p) \to k$, $\theta(d^{-1}) \to k$, $pS(x) \to p(s_1 + \cdots + s_v) = ps$ and $T \to v$. Applying this homomorphism to (3.14) gives

$$k^2 = ps + \lambda v \tag{3.15}$$

but as $k^2 - k = \lambda(v - 1) = \lambda v - \lambda$, it follows that

$$ps = k^2 - \lambda v = k - \lambda = n. \tag{3.16}$$

Applying the automorphism $x \to x^p$ of G to (3.7) gives

$$\theta(d^p)\theta(d^{-p}) = n + \lambda T. \tag{3.17}$$

Applying the automorphism $x \to x^{-1}$ of G to (3.14) gives

$$\theta(d)\theta(d^{-p}) = pS(x^{-1}) + \lambda T. \tag{3.18}$$

The product of the left hand sides of (3.7) and (3.17) is the same as that for (3.14) and (3.18) so that equating the products of the right hand sides gives

$$(pS(x) + \lambda T)(pS(x^{-1}) + \lambda T) = (n + \lambda T)^2 \tag{3.19}$$

Now from (3.16)

$$pS(x)T = pS(x^{-1})T = psT = nT. \tag{3.20}$$

Thus (3.19) reduces to

$$p^2 S(x) S(x^{-1}) = n^2. \tag{3.21}$$

So far we have not used the condition (iii) $p > \lambda$. Here (3.14) takes the form

$$a_1 x_1 + a_2 x_2 + \cdots + a_\nu x_\nu$$
$$= p(s_1 x_1 + \cdots + s_\nu x_\nu) + \lambda(x_1 + \cdots + x_\nu) \tag{3.22}$$

where the a_i are non-negative integers, so that in $a_i = ps_i + \lambda$, $i = 1, \ldots, \nu$, as $p > \lambda$ if follows that $s_i \geq 0$, $i = 1, \ldots, \nu$. But then (3.21) has the form

$$p^2(s_1 x_1 + \cdots + s_\nu x_\nu)(s_1 x_1^{-1} + \cdots + s_\nu x_\nu^{-1}) = n^2. \tag{3.23}$$

On the left we cannot have a non-zero term $s_i s_j x_i x_j^{-1}$ since all coefficients are non-negative and on the right only the identity has a non-zero coefficient. Hence $S(x)$ consists of a single term sx and $pS(x) = psx = nx$. Thus (3.14) takes the simpler form

$$\theta(d^p)\theta(d^{-1}) = nx + \lambda T. \tag{3.24}$$

Multiplying (3.24) by $\theta(d)$ gives

$$\theta(d^p)\theta(d^{-1})\theta(d) = nx\theta(d) + \lambda\theta(d) + \lambda\theta(d)T. \tag{3.25}$$

Using (3.7) we have

$$\theta(d^p)(n + \lambda T) = nx\theta(d) + \lambda\theta(d)T. \qquad (3.26)$$

Now $\theta(d)T = kT = \theta(d^p)T$ so that we have

$$n\theta(d^p) = nx\theta(d) \qquad (3.27)$$

and dividing by n

$$\theta(d^p) = x\theta(d). \qquad (3.28)$$

But this equation says that the sets $\{d_1^p, d_2^p, \ldots, d_k^p\}$ and $\{xd_1, xd_2, \ldots, xd_k\}$ are the same and this is the multiplicative form of the relation (3.3) saying that p is a multiplier. This completes the proof of the theorem.

The condition $(p, v) = 1$ is clearly necessary for p to be a multipler and the condition that p divides $n = k - \lambda$ is a reasonable source for the multiplier. But the condition $p > \lambda$ is a troublesome one, since although it is used in the proof, no example is known where this condition is necessary.

THEOREM 3.3 (Parker): *An automorphism of a symmetric block design fixes the same number of blocks and points.*

Proof: If a block design has v points $x_1, \ldots, x_v$ and b blocks $B_1, \ldots, B_b$ we may define the incidence matrix A by

$$A = [a_{ij}] \; i = 1, \ldots, v, j = 1, \ldots, b,$$
$$a_{ij} = 1, \text{ if } x_i \in B_j, \qquad (3.29)$$
$$a_{ij} = 0, \text{ if } x_i \notin B_j.$$

The properties of this incidence matrix are considered in some detail in the article by H. J. Ryser in this volume. If the parameters of the design D are v, b, r, k, λ then

$$AA^T = (r - \lambda)I + \lambda J, \qquad (3.30)$$

where I is the identity matrix of order v and J the matrix of order v all of whose entries are 1's. In particular Ryser shows that if the design is symmetric and $b = v$, then the matrix A is non-singular. If α is an automorphism of the symmetric design D, let P be the matrix representing the permutation α induces on the points of D and Q that for the blocks. Then the fact that α is an automorphism is expressed by the matrix relation

$$P^{-1}AQ = A. \qquad (3.31)$$

From this and the fact that A is non-singular it follows that

$$Q = APA^{-1}. \qquad (3.32)$$

Hence taking traces on both sides of this equation it follows that

$$\mathrm{tr}(Q) = \mathrm{tr}(P). \qquad (3.33)$$

But $\mathrm{tr}(Q)$ is the number of blocks fixed by α and $\mathrm{tr}(P)$ the number of points fixed by α. This equality is the statement of the theorem.

Since a multiplier of a difference set always fixes the zero element of the additive group whose elements are the points of the design, this theorem tells us that a multiplier must also fix a block. For example with $v = 11$, $k = 5$, $\lambda = 2$ since $k - \lambda = 3$ Theorem 3.2 tells us that $p = 3$ is a multiplier. If B is a block fixed by this multiplier, then if c is a non-zero point on this block we must also have $c, 3c, 9c, 27c = 5c, 15c = 4c$ as points of this block. Hence apart from the factor c, the block can only be 1, 3, 4, 5, 9 (mod 11). This yields the cyclic symmetric design of (2.8).

The multiplier theorem is very useful in the construction of difference sets or in showing that they do not exist. A number of infinite classes of difference sets, most of which depend on arithmetical properties of residues modulo p or of elements in a finite field $\mathrm{GF}(q)$, are known. The main work on this subject is L. Baumert's book [5] "Cyclic difference sets".

TYPE S (Singer difference sets [32]): These are hyperplanes in the n-dimensional projective geometry $\mathrm{PG}(n, q)$ over $\mathrm{GF}(q)$. The parameters are

$$v = \frac{q^{n+1} - 1}{q - 1}, k = \frac{q^n - 1}{q - 1}, \lambda = \frac{q^{n-1} - 1}{q - 1}.$$

TYPE Q (Quadratic residues in GF(q), $q \equiv 3 \pmod 4$):

$$v = q = p^r = 4t - 1, k = 2t - 1, \lambda = 2t - 1.$$

TYPE H_6 (p is a prime of the form $p = 4x^2 + 27$): There will exist a primitive root r modulo p such that $\text{Ind}_r(3) \equiv 1 \pmod 6$. The $(p - 1)/2$ residues a_i such that $\text{Ind}_r(a_i) \equiv 0, 1,$ or $3 \pmod 6$ will form a difference set with parameters

$$v = p = 4t - 1, k = 2t - 1, \lambda = t - 1.$$

TYPE T (Twin primes): Let p and $q = p + 2$ be primes. Let r be a number such that r is a primitive root of p and also of q. Then $r^i \pmod{pq}$ $i = 1, \ldots, (p - 1)(q - 1)/2$, and $0, q, \ldots, (p - 1)q \pmod{pq}$ form a difference set with $v = pq = 4t - 1, k = 2t - 1, \lambda = t - 1$.

TYPE B (Biquadratic residues of primes $p = 4x^2 + 1$, x odd): Here $v = p = 4x^2 + 1, k = x^2, \lambda = (x^2 - 1)/4$.

TYPE B_0 (Biquadratic residues and zero modulo primes $p = 4x^2 + 9$, x odd): Here $v = 4x^2 + 9, k = x^2 + 3, \lambda = (x^2 + 3)/4$.

TYPE O (Octic residues of primes $p = 8a^2 + 1 = 64b^2 + 9$ with odd a, b odd): Here $v = p, k = a^2, \lambda = b^2$.

TYPE O_o (Octic residues and zero for primes $p = 8a^2 + 49 = 64b^2 + 441$, a odd, b even): Here $v = p, k = a^2 + 6, \lambda = b^2 + 7$.

TYPE W_4 (A generalization of T developed by Whiteman [38]): Let p be a prime $p \equiv 1 \pmod 4$ and let $q = 3p + 2$ also be a prime. Suppose also that $pq = v = 1 + 4x^2$ with x odd. Then take r to be a primitive root of both p and q. Writing $d = (p - 1)(q - 1)/4$, the residues $1, r, r^2, \ldots, r^{d-1}, 0, q, 2q, \ldots, (p - 1)q \pmod{pq}$ are a difference set with $v = pq, k = (v - 1)/4, \lambda = (v - 5)/16$.

TYPE GMW (Gordon, Mills, Welch [12]): The parameters are the same as those of the Singer type

$$v = \frac{q^{n+1} - 1}{q - 1}, k = \frac{q^n - 1}{q - 1}, \lambda = \frac{q^{n-1} - 1}{q - 1}.$$

Here if we can write $n + 1$ in the form $n + 1 = mM$ with $m \geq 3$ and if M is the product of r prime numbers, not necessarily distinct, then there are at least 2^r inequivalent difference sets with these parameters.

We write (x, y) mod (r, s) for elements (x, y) in the direct product of the cyclic group of order r and that of order s, taking x modulo r and y modulo s. A symmetric design with $v = b = 36$, $r = k = 15$, $\lambda = 6$ is given by the base block of 15 points modulo $(6, 6)$:

$$\begin{array}{ccccc}
(0,1) & (0,2) & (0,3) & (0,4) & (0,5) \\
(1,0) & (2,0) & (3,0) & (4,0) & (5,0) \\
(1,1) & (2,2) & (3,3) & (4,4) & (5,5).
\end{array} \quad (3.34)$$

This is an example of a difference set on a non-cyclic Abelian group where the multiplier theorem does not give a multiplier. $t = -1$ is however a multiplier. But every known cyclic difference set has a multiplier.

A group may move the points of a design in more than one orbit and also the blocks in more than one orbit. The first major investigation of such designs was made by R. C. Bose [6] in 1939. A number of methods are given in the writer's book [14].

R. M. Wilson [40] has found constructions for block designs from finite fields $F = \text{GF}(q)$ which have the additive group A of $\text{GF}(q)$ as automorphisms. Let e be a divisor of $q - 1$ and let H be the multiplicative group of the $q - 1$ elements different from 0. H is a cyclic group and the eth powers of elements of H form a subgroup H^e of order $(q - 1)/e$. A generator w of H is called a primitive root of $\text{GF}(q)$ and the elements $H_m{}^e = \{w^t | t \equiv m \pmod{e}\}$ form a coset of $H^e = H_0{}^e$. The cosets are $H_0{}^e, H_1{}^e, \ldots, H_{e-1}{}^e$. Given a set $B = \{b_1, \ldots, b_k\}$ of k distinct elements of F we say the elements of B are *evenly distributed* over the eth power cosets if and only if there are the same number s of differences $b_i - b_j$ in each of the cosets $H_m{}^e$. Here necessarily $k(k - 1) = se$. As an example take

$$B = \{0, 1, 3, 24\} \mod 37. \quad (3.35)$$

Here 2 is a primitive root and with $e = 6H^6$ consists of the residues 1, 10, 11, 26, 27, 36. Here the 12 differences of B are evenly distributed with respect to H^6. Here $e/2 = 3$ and the solutions of $x^3 = 1$ in H are 1, 10, 26, and multiplying B by these three residues we have three blocks

$$\{0, 1, 3, 24\} \quad \{0, 10, 18, 30\} \quad \{0, 4, 26, 32\}. \tag{3.36}$$

With the additive group modulo 37 as automorphism group these three blocks determine a design with $v = 37$, $b = 111$, $r = 12$, $k = 4$, $\lambda = 1$. Wilson's main result shows that blocks with evenly distributed differences exist if q is sufficiently large compared to k. This is a consequence of a more general result, which will be described briefly. Let $(a_1, a_2, \ldots, a_r)$ be an ordered set of r elements from $GF(q)$. We say that $(a_1, \ldots, a_r)$ is consistent with a choice $C = \{m_{ij}\}$ of eth power residues providing that for all values $1 \le i < j \le r$, $a_i - a_j \in H_{m_{ij}}{}^e$. He proves that the number of consistent sets for each choice is sufficiently near the average so that if q is sufficiently large, there is a consistent r-set for every choice. Let $N(C)$ be the number of consistent r-sets for a choice $C = \{m_{ij}\}$. Two theorems describe his results:

THEOREM 3.4: *Let* $q \equiv 1 \pmod{e}$ *be a prime power. Then for any choice* $C = \{m_{ij}\}$ $1 \le i < j \le r$

$$\left| N(C) - \frac{q(q-1) \cdots (q-r+1)}{e^{\frac{1}{2}r(r-1)}} \right| < e^{\frac{1}{2}r - \frac{1}{2}} q^{r - \frac{1}{2}}.$$

THEOREM 3.5: *Let q be a prime power, $q > \left|\frac{1}{2}k(k-1)\right|^{k(k-1)}$. There exists a design with $v = q$, k, and λ with base blocks with respect to A, the additive group of $GF(q)$ if and only if $\lambda(q - 1) \equiv 0 \pmod{k(k-1)}$.*

For example, with $k = 6$, $\lambda = 1$ and if $q \equiv 1 \pmod{30}$ there is a design with $v = q$ provided that $q > 15^{30}$. This lower bound appears to be much too high, and a sharpening of the inequality in Theorem 3.4 would improve this.

There are relations between graphs and designs. Here a graph $\mathcal{G}$ is understood to consist of a set of v points called vertices and a set of undirected arcs joining certain distinct pairs of points. A graph is called *regular* if every vertex is joined to the same number $k < v - 1$ of vertices. A regular graph is called *strongly regular* if for any two distinct vertices P and Q the number of vertices joined to both P and Q depends only on whether P and Q are joined or not there being λ joined to both if P and Q are joined and μ joined to both if P and Q are not joined.

A graph is determined by its incidence matrix A where the points are $P_1, \ldots, P_v$:

$$\begin{aligned}
&A = [a_{ij}], \ i, j = 1, \ldots, v, \\
&a_{ii} = 0, \\
&a_{ij} = 1, \text{ if } P_i \text{ and } P_j \text{ are joined}, \\
&a_{ij} = 0, \text{ if } P_i \text{ and } P_j \text{ are not joined}.
\end{aligned} \quad (3.37)$$

As usual we write J for the matrix consisting entirely of 1's. Let us define the matrix B by putting $B = J - I - A$, so that B would be the matrix of the complementary graph in which distinct points P_i and P_j are joined if and only if they are not joined in the original graph. In a strongly regular graph the number l of points not joined to a given point is determined by $l = v - 1 - k$. Then the incidence matrix A of a strongly regular graph satisfies the following relations:

$$AJ = JA = kJ,$$

$$I + A + B = J,$$

$$AA^T = kI + \lambda A + \mu B, \quad (3.38)$$

$$A = A^T,$$

$$(A - kI)(A^2 - (\lambda - \mu)A - (k - \mu)I) = 0.$$

If the graph is connected, as we shall assume, then A has the eigenvalue k with multiplicity one. The other two eigenvalues s and t are given by

$$\begin{Bmatrix} s \\ t \end{Bmatrix} = \frac{\lambda - \mu \pm \sqrt{d}}{2}, \quad d = (\lambda - \mu)^2 + 4(k - \mu). \quad (3.39)$$

Let their multiplicities be f_2 and f_3. Then we find, using the fact that A has trace 0, and that the eigenvalue k has multiplicity one, that

$$\begin{Bmatrix} f_2 \\ f_3 \end{Bmatrix} = \frac{2k + (\lambda - \mu)(k + l) \pm \sqrt{d}\,(k + l)}{\pm 2\sqrt{d}}. \quad (3.40)$$

We also have the relation

$$\mu l = k(k - \lambda - 1) \quad (3.41)$$

and, as we assume our graph connected, then $\mu > 0$. We may conclude that one of the two following cases arises:

Case I. $k = l$, $\mu = \lambda + 1 = k/2$ and $f_2 = f_3 = k$.
Case II. $d = (\lambda - \mu)^2 + 4(k - \mu)$ is a square and the multiplicities f_2, f_3 in (3.40) are positive integers.

In a strongly regular graph if $\lambda = \mu$, then the sets of k points joined to P_i, $i = 1, \ldots, v$, form the blocks of a symmetric design. If $\lambda = \mu + 2$, then the sets of $k^* = k + 1$ points consisting of the k points joined to P_i together with P_i itself for $i = 1, \ldots, v$ form the blocks of a symmetric design.

If n is even and we are given $n/2$ families of parallel lines on n^2 points, then form a graph by joining each point to the other points on a line with it. This gives a strongly regular graph with $\lambda = \mu$ and a symmetric design with parameters

$$v = n^2, k = n(n - 1)/2, \lambda = n(n - 2)/4.$$

This we have if there are $\frac{1}{2}n - 2$ mutually orthogonal Latin squares of order n. Thus a single Latin square of order 6 yields a symmetric design with parameters (36, 15, 6).

A transitive permutation group G and v points is said to be a rank 3 group if the stabilizer of a point a has exactly three orbits (a), $\Delta(a)$, $\Gamma(a)$, where we take $k = |\Delta(a)|$, $l = |\Gamma(a)|$. Joining the point a to every point of the orbit $\Delta(a)$ gives a strongly regular graph if G is of even order. In a number of cases rank 3 groups yield symmetric block designs. Here we quote one interesting case [17] which gives the only known symmetric design parameters (56, 11, 2). The group is $PSL_3(4) = G$, a group of collineations of the projective plane of order 4 represented on 56 points. Here $G = \langle a, d \rangle$ where

$$a = (1, 2, 3, 4, 5, 6, 7)(8, 9, 10, 11, 12, 13, 14)$$

$$(15, 16, 17, 18, 19, 20, 21)(22, 23, 24, 25, 26, 27, 28)$$

$$(29, 30, 31, 32, 33, 34, 35)(36, 37, 38, 39, 40, 41, 42)$$

$$(43, 44, 45, 46, 47, 48, 49)(50, 51, 52, 53, 54, 55, 56).$$
(3.42)
$$d = (1)(2, 34)(3, 54)(4, 39)(5, 13)(6, 29)(7, 56)(8)(9, 44)$$

$$(10, 16)(11)(12, 19)(14)(15, 41)(17, 55)(18, 52)(20, 42)(21, 24)(22, 26)$$

$$(23)(25)(27, 36)(28, 40)(30, 47)(31, 33)(32, 50)(35, 43)(37, 45)$$

$$(38)(46, 53)(48)(49, 51).$$

Here G is of order 20, 160 and is rank 3 with $v = 56$, $k = 10$, $l = 45$, $\lambda = 0$, $\mu = 2$. The point 1 and the points of the 10 orbit in G_1 form a block

$$B: \{1, 12, 19, 23, 30, 37, 45, 47, 48, 49, 51\}. \quad (3.43)$$

The images of B_1 under G form a symmetric block design with $v = 56$, $k = 11$, $\lambda = 2$.

4. RECURSIVE CONSTRUCTIONS. TWO FAMOUS CONJECTURES

Orthogonal Latin squares were described in section 2. A natural question is to ask how many mutually orthogonal Latin squares of order n there can be. If the maximum number is denoted by $N(n)$, then it is trivial that $N(n) \le n - 1$ in every case and as was observed in section 2, for a prime power $q = p^r$ we do have $N(q) = q - 1$. Easy arguments show that $N(n) \ge 2$ whenever n is odd (and of course $n \ge 3$) n is a multiple of 4. Here $N(2) = 1$ trivially, and in 1900 Tarry [34] showed by a complete enumeration of cases that $N(6) = 1$. Indeed it appeared that $N(4t + 2) = 1$ and this assertion was the substance of a statement made by Euler in 1782. At this time a pair of orthogonal squares was described as a Graeco-Latin square, consisting of two superimposed orthogonal Latin squares, one in Latin letters, the other in Greek letters. Euler's statement was "I do not hesitate to conclude that it is impossible to produce any complete (Graeco-Latin) square of 36 entries, and the same impossibility extends to the cases of $n = 10$, $n = 14$, and in general to all unevenly even numbers".

EULER CONJECTURE: *There do not exist two orthogonal squares of order n when n is of the form $n = 4t + 2$.*

This conjecture was completely demolished in 1959 by joint work of R. C. Bose, Ernest Parker, and S. S. Shrikhande [11, 12, 27] who succeeded in showing by recursive constructions that $N(n) \ge 2$ for every n including those of the form $n = 4t + 2$ except for the values $n = 2$ and $n = 6$.

It was noted in section 2 that the parameters v, b, r, k, λ of a block design satisfy the relations

$$bk = vr, \qquad r(k - 1) = \lambda(v - 1). \tag{4.1}$$

We may rewrite these relations in the form

$$r = \lambda(v - 1)/(k - 1),$$
$$b = \lambda v(v - 1)/(k(k - 1), \tag{4.2}$$

so that $v, k,$ and λ determine r and b. Here (4.2) is equivalent to congruences on v, k, λ namely

$$\lambda(v - 1) \equiv 0 \pmod{k - 1},$$
$$\lambda v(v - 1) \equiv 0 \pmod{k(k - 1)}. \tag{4.3}$$

It is reasonable to conjecture that the necessary conditions (4.3) are also sufficient if v is sufficiently large.

EXISTENCE CONJECTURE FOR BALANCED INCOMPLETE BLOCK DESIGNS. *Given positive integers k and λ, balanced incomplete block designs on v points with block size k and index λ exist for all sufficiently large integers v satisfying the congruences $\lambda(v - 1) \equiv 0 \pmod{k - 1}$ and $\lambda v(v - 1) \equiv 0 \pmod{k(k - 1)}$.*

Using recursive constructions, progress was made on this conjecture by H. Hanani [18, 20] and a complete proof of its correctness given by R. M. Wilson [39] in 1973. Wilson proved even more, proving the analogue of this conjecture for pairwise balanced designs.

The Kronecker product $H \times K$ defined in (2.19) yields a recursive construction for combinatorial systems in a number of cases.

THEOREM 4.1 (MacNeish [23]): *If there are t mutually orthogonal Latin squares of order m and also t mutually orthogonal squares of order s, then there are t mutually orthogonal squares of order ms.*

Let $L_1, L_2, \ldots, L_t$ be t mutually orthogonal squares of order m whose entries are the indeterminates $x_1, x_2, \ldots, x_m$ and let $K_1, K_2, \ldots, K_t$ be mutually orthogonal squares of order s whose entries are the indeterminates $y_1, \ldots, y_s$. Then treating these as matrices the t matrices $L_1 \times K_1, L_2 \times K_2, \ldots, L_t \times K_t$ can easily be shown to be t mutually orthogonal squares of order ms in the ms quantities $x_i y_j$.

COROLLARY: *If $n = p_1^{e_1} p_2^{e_2} \cdots p_r^{e_r}$, where $p_1, \ldots, p_r$ are distinct primes, then $N(n) \geq \min(p_i^{e_i} - 1)$ $i = 1, \ldots, r$.*

THEOREM 4.2: *If there are Steiner triple systems with v_1 points and with v_2 points, then there is a Steiner triple system with $v_1 v_2$ points.*

If $x_1, \ldots, x_v$ are indeterminates corresponding to the points of a Steiner triple system $S(v)$, we define a matrix $A = [a_{ij}]$ associated with this system putting $a_{ii} = x_i$, and if $i \neq j$ and if x_i, x_j, x_k correspond to a triple we put $a_{ij} = x_k$. It is not difficult to show that if $S(v_1)$ corresponds to a matrix A_1 on indeterminates $x_1, \ldots, x_{v_1}$ and $S(v_2)$ to A_2 on $y_1, \ldots, y_{v_2}$, then the Kronecker product $A_1 \times A_2$ is easily shown to correspond to a Steiner triple system with $v_1 v_2$ points.

THEOREM 4.3: *If there are Hadamard matrices of orders m and n then there is a Hadamard matrix of order mn.*

It has been observed in section 2 that the Kronecker product of an H_m and an H_n is an H_{mn}.

The starting point for the construction of mutually orthogonal Latin squares is naturally the system of affine planes of order $q = p^r$, a prime power, for which the number of squares is the maximum, $N(q) = q - 1$. These lead immediately to the bound given in the corollary to Theorem 4.1, which MacNeish conjectured was best possible. His conjecture is of course even stronger than the Euler conjecture. In a series of papers [11, 12, 27] R. C. Bose, E. T. Parker, and S. S. Shrikhande succeeded in proving $N(n) \geq 2$ for every n except $n = 2$ and 6 for which the result is false. Their constructions involved an interplay between pairwise balanced designs and mutually orthogonal squares. A main theorem is:

THEOREM 4.4 (Bose, Parker, Shrikhande): *Suppose there is a PBD$(v, K, 1)$ where $K = \{k_1, \ldots, k_r, k_{r+1}, \ldots, k_m\}$ and suppose for k_i, $i = 1, \ldots, r$ the blocks of size k_i are disjoint. Then $N(v) \geq \min(N(k_1), \ldots, N(k_r), N(k_{r+1}) - 1, \ldots, N(k_m) - 1)$.*

With this theorem and a few similar theorems it was possible to prove $N(v) \geq 2$ for $v \geq 730$ and along with special constructions for $v = 12 + 10$ and $v = 14, 26, 38$, it was shown that $N(v) \geq 2$ for all $v > 6$ completely disproving the Euler conjecture. A detailed account of this work is contained in Chapter 13 of the writer's book [14].

There have been a number of improvements on this result.

THEOREM 4.5 (Chowla, Erdös and Straus [9]): $N(n) > 1/3n^{1/91}$ *for sufficiently large n.*

THEOREM 4.6 (K. Rogers [30]): $N(n) > n^{1/42+\epsilon}$ *for* $n > n_\epsilon$.

THEOREM 4.7 (R. Wilson [41]): $N(n) > n^{1/17-2}$ *for large n. Also* $N(n) \geq 6$ *whenever* $n > 90$.

THEOREM 4.8 (H. Hanani [19]): $N(n) \geq 3$ *for* $n > 51$, $N(n) \geq 5$ *for* $n > 62$ *and* $N(n) \geq 29$ *for* $n > 34,115,553$.

Given a set K of positive integers we denote by $B(K, \lambda)$ the set of those v's for which a pairwise balanced design $D(v, K, \lambda)$ exists. A recursive theorem due to Hanani [18] is the following:

THEOREM 4.9: *If* $n \in B(K, \lambda)$ *and if for every* $k_i \in K$, $(k - 1)k_i + 1 \in B(k, 1)$ *holds, then* $(k - 1)n + 1 \in B(k, \lambda)$.

This and similar constructions led to the following result:

THEOREM 4.10 (H. Hanani [18, 20]): *For* $k = 3, 4, 5$ *and any* λ *the necessary conditions* $\lambda(v - 1) \equiv 0 \pmod{k - 1}$ *and* $\lambda v(v - 1) \equiv 0 \pmod{k(k - 1)}$ *are sufficient for the existence of a balanced incomplete block design* $D(v, k, \lambda)$ *with the single exception* $v = 15$, $k = 5$, $\lambda = 2$ *for which a design does not exist.*

Richard Wilson [39, 40] developed a number of constructions in his proof of the truth of the existence conjecture for balanced incomplete block designs and pairwise balanced designs. One of his methods can be considered a major generalization of a construction for Steiner triple systems given by E. H. Moore [26] in 1893. The Moore construction will be given here and from it a partial description of Wilson's method. Write $S(v)$ for a Steiner triple system $(v, 3, 1)$.

THEOREM 4.11 (E. H. Moore): *If there is a Steiner triple system $S(v_2)$ containing a subsystem $S(v_3)$ where we permit $v_3 = 0, 1$, and there is an $S(v_1)$, then there is an $S(v)$ with $v = v_3 + v_1(v_2 - v_3)$.*

Proof: We construct a set S of $v = v_3 + v_1(v_2 - v_3)$ points listed in $v_1 + 1$ sets, writing $s = v_2 - v_3$:

$$S_0: a_1, \ldots, a_{v_3},$$
$$S_1: b_{11}, \ldots, b_{1s},$$
$$\ldots\ldots\ldots\ldots$$
$$S_i: b_{i1}, \ldots, b_{is},$$
$$\ldots\ldots\ldots\ldots$$
$$S_{v_1}: b_{v_11}, \ldots, b_{v_1s}.$$

We use three rules to construct triples from S:

1. We associate the points $a_1, \ldots, a_{v_3}$ of S_0 with the points of the given $S(v_3)$ and take the triples (a_i, a_j, a_k) as triples of S.
2. We make S_0 and S_i, $i = 1, \ldots, v_1$ correspond to the $S(v_2)$ identifying the points of S_0 with those of the subsystem $S(v_3)$. From $S(v_2)$ we obtain triples (a_i, a_j, a_k) as in 1 above and further triples (a_j, b_{ir}, b_{it}) and (b_{iu}, b_{iv}, b_{iw}).
3. Writing the system $S(v_1)$ on the numbers $1, \ldots, v_1$ if (j, k, r) is a triple of $S(v_1)$ and if $x + y + z \equiv 0 \pmod{s}$, we take as a triple of S (b_{jx}, b_{ky}, b_{rz}).

It is not difficult to verify the correctness of this construction. Let us turn to an interpretation of this construction in terms of Wilson's generalization. Let X be the triple system with $v = v_3 + v_1(v_2 - v_3)$ points just given and let Y be the system with v_1

points $\{1, 2, \ldots, v_1\}$. We define a *morphism* from X to Y and write this as $X \to Y$. The morphism is a partial mapping f of points of X onto points of Y. In this case the mapping f is $b_{ij} \to i$ for $i = 1, \ldots, v_1$ and f maps the points $a_1, \ldots, a_{v_3}$ of S_0 onto the void set. The number d of points mapped onto the void set is called the deficiency. Here $d = v_3$. In the morphism it is further required that for the points of a block A, in the mapping $A \to f(A)$, that either (i) $|f(A)| \le 1$ or (ii) $A \to f(A)$ is one-to-one and there is a block B of Y such that $f(A) \subseteq B$.

In the Moore construction we have $|f(A)| = 0$ for blocks constructed by the first rule, $|f(A)| = 1$ for blocks constructed by the second rule, and under the third rule $f(A) = b$, a block of Y.

In the morphism f, $X \to Y$, Wilson calls X the pre-image of Y under f. He supposes that Y is given and wishes to find f and a pre-image X. For each point y of Y, we have a number $w(y)$, the weight of y, where under f the number of points mapped onto y (or the void set) is $w(y) + d$, so that this is the cardinality of $f^{-1}(y)$.

A first necessary condition is for each point y of Y that $w(y) + d$ be the cardinality of a PBD containing a subsystem of cardinality d. A second necessary condition is that for every block B of Y with points $y_1, \ldots, y_m$ there is a group divisible design with groups of sizes $w(y_1), w(y_2), \ldots, w(y_m)$ and block sizes from K. In the Moore construction $w(y) = s = v_2 - v_3$ for every y and for a block B with three points (j, k, r) the third rule describes a group divisible design with three groups (S_j, S_k, S_r) of size s and block sizes 3. It is in fact the transversal design for a Latin square of size s by s.

For pairwise balanced designs with $\lambda = 1$ and block sizes from K, Wilson shows that the two necessary conditions above are also sufficient for the existence of the pre-image X and the morphism f. He also shows that this construction can be carried out so that X has a subsystem isomorphic to Y. In the Moore construction the points $\{b_{1s}, \ldots, b_{vs}\}$ form an $S(v_1)$.

This description of one of Wilson's constructions gives the flavor of much of his work. Given a set $K = \{k_1, k_2, \ldots,\}$ of positive integers, define $\alpha(K)$ to be the greatest common divisor of the $k_i - 1$, $k_i \in K$ and $\beta(K)$ to be the greatest common divisor of all $k_i(k_i - 1)$, $k_i \in K$. Starting from the designs whose existence is guaranteed by the result [40] quoted in Theorem 3.5 he uses a large variety of recursive constructions to prove the main theorem.

THEOREM 4.12 (Wilson [39]): *Given K, a set of positive integers, and λ, there will exist a pairwise balanced design D(v, K, λ) if the necessary congruences $\lambda(v - 1) \equiv 0 \pmod{\alpha(K)}$ and $\lambda v(v - 1) \equiv 0 \pmod{\beta(K)}$ are satisfied providing that v is sufficiently large.*

Note that this theorem includes the existence conjecture as a special case, when K consists of a single number k. But in contrast with Hanani's results given in Theorem 4.10, the lower bounds implied in the "sufficiently large" though computable, are very large indeed.

Unfortunately there is not space here to describe further modes of construction which have been pursued vigorously recently. One of these is the construction of Hadamard matrices in which the most significant steps have been taken independently by R. Turyn [35] and J. Wallis [37]. Another is the construction of t-designs. A t-design is a set of points, the blocks being of uniform size k chosen from v points, with the property that every subset of $t \geq 2$ points occurs the same number of times, λ, in the set of blocks. Here if $t = 2$ the design is an ordinary block design. Assmus and Mattson [3, 4] and V. Pless [28] have applied the principles of coding theory [36] in this construction. W. O. Alltop [1] has also made important contributions.

REFERENCES

1. W. O. Alltop, "An infinite class of 5-designs", *J. Combinatorial Theory,* **12** (1972), 390-395.
2. R. H. Bruck, "Difference sets in a finite group", *Trans. Amer. Math. Soc.,* **78** (1955), 464-481.
3. E. F. Assmus and H. F. Mattson, "Perfect codes and the Mathieu groups", *Arch. Math.,* **17** (1966), 121-135.
4. _____, "New 5-designs", *J. Combinatorial Theory,* **6** (1969), 122-151.
5. L. D. Baumert, "Cyclic Difference Sets", *Lecture Notes in Mathematics No. 182,* Springer-Verlag, Berlin-Heidelberg-New York, 1971.
6. R. C. Bose, "On the construction of balanced incomplete block designs", *Ann. Eugenics,* **9** (1939), 353-399.

7. R. C. Bose, E. T. Parker, and S. Shrikhande, "Further results on the construction of mutually orthogonal Latin squares and the falsity of Euler's conjecture", *Canad. J. Math.*, **12** (1960), 189-203.
8. R. C. Bose and S. Shrikhande, "On the construction of sets of mutually orthogonal Latin squares and the falsity of a conjecture of Euler", *Trans. Amer. Math. Soc.*, **95** (1960), 191-209.
9. S. Chowla, P. Erdös, and E. G. Straus, "On the maximal number of pairwise orthogonal squares of a given order", *Canad. J. Math.*, **12** (1960), 204-208.
10. P. Dembowski, *Finite Geometries*, Springer-Verlag, Berlin-Heidelberg-New York, 1968.
11. R. A. Fisher, "An examination of the different possible solutions of a problem in incomplete blocks", *Ann. Eugenics*, **10** (1940), 52-75.
12. B. Gordon, W. H. Mills, and L. R. Welch, "Some new difference sets", *Canad. J. Math.*, **14** (1962), 614-625.
13. Marshall Hall, Jr., "Note on the Mathieu group M_{12}", *Arch. Math.*, **13** (1962), 334-340.
14. ____, *Combinatorial Theory*, Blaisdell, Waltham, Mass., 1967.
15. ____, "Designs with transitive automorphism groups", *Amer. Math. Soc. Proc. of Symposia in Pure Mathematics*, **19** (1971), 109-113.
16. M. Hall, Jr., and H. J. Ryser, "Cyclic incidence matrices", *Canad. J. Math.*, **3** (1959), 495-502.
17. M. Hall, R. Lane, and D. Wales, "Designs derived from permutation groups", *J. Combinatorial Theory*, **8** (1970),, 12-22.
18. H. Hanani, "The existence and construction of balanced incomplete block designs", *Ann. Math. Statist.*, **32** (1961), 361-386.
19. ____, "On the number of orthogonal Latin squares", *J. Combinatorial Theory*, **8** (1970), 247-271.
20. ____, "On balanced incomplete block designs with blocks having five elements", *J. Combinatorial Theory*, **12** (1972), 184-201.
21. D. G. Higman, "Finite permutation groups of rank 3", *Math Z.*, **86** (1967), 145-156.
22. T. P. Kirkman, "On a problem in combinations", *Cambridge and Dublin Math. J.*, **2** (1847), 191-204.
23. H. F. MacNeish, "Euler squares", *Ann. of Math.*, **23** (1922), 221-227.
24. F. J. MacWilliams, "A theorem on the distribution of weights in a systematic code", *Bell System Tech. J.*, **42** (1963), 79-94.
25. H. B. Mann, *Analysis and Design of Experiments*, Dover, New York, 1949.
26. E. H. Moore, "Concerning triple systems", *Math. Ann.*, **43** (1893), 271-285.
27. E. T. Parker, "Construction of some sets of mutually orthogonal Latin squares", *Proc. Amer. Math. Soc.*, **10** (1959), 946-949.
28. Vera Pless, "Symmetry codes over GF(3) and new five-designs", *J. Combinatorial Theory*, **12** (1972), 119-142.
29. D. K. Ray-Chaudhuri and R. M. Wilson, "Solution of Kirkman's schoolgirl problem", Combinatorics, *Amer. Math. Soc. Proc. Pure Math.*, **19** (1971), 187-203.

30. K. Rogers, "A note on orthogonal Latin squares", *Pacific J. Math.*, **114** (1964), 1395-1397.
31. H. J. Ryser, "Combinatorial mathematics", *Carus Monograph 14*, Mathematical Association of America, 1963.
32. J. Singer, "A theorem in finite projective geometry and some applications to number theory", *Trans. Amer. Math. Soc.*, **43** (1938), 377-385.
33. J. Steiner, "Combinatorische aufgabe", *J. Reine Angew. Math.*, **45** (1853), 181-182.
34. G. Tarry. "Le problème des 36 officiers", *C. R. Assoc. Fr. Av. Sci.*, **1** (1900), 122-123; **2** (1901), 170-203.
35. R. Turyn, "Hadamard matrices, Buamert-Hall units, four symbol sequences, pulse compression and surface wave enclodings", *J. Combinatorial Theory Ser. A*, **16** (1974), 313-333.
36. J. H. van Lint, "Coding theory", Lecture Notes in Mathematics, No. 201, Springer-Verlag, Berlin-Heidelberg-New York, 1971.
37. J. Wallis, "Hadamard matrices of order $28m$, $36m$, and $44m$", *J. Combinatorial Theory Ser. A*, **15** (1973), 323-328.
38. A. L. Whiteman, "A family of difference sets", *Illinois J. Math.*, **6** (1962), 107-121.
39. R. M. Wilson, "An existence theory for pairwise balanced designs I, Composition Theorems and Morphisms", *J. Combinatorial Theory*, **13** (1972), 220-245. II. "The structure of PBD-closed sets and the existence conjectures", *J. Combinatorial Theory*, **13** (1972), 246-273. III. "Proof of the existence conjectures", *J. Combinatorial Theory Ser. A*, **18** (1975), 71-79.
40. ———, "Cyclotomy and difference families in elementary Abelian groups", *J. Number Theory*, **4** (1972), 17-47.
41. ———, "Concerning the number of mutually orthogonal latin squares", *Discrete Math.*, **9** (1974), 181-198.

AUTHOR INDEX

Alder, H. L., 140
Alltop, W. O., 251
Alter, R., 140
Amitsur, S. A., 7, 19
Anand, H., 139, 140
Anderson, I., 76
Andrews, G. E., 140, 141
Assmus, E. F., 251

Baker, K., 36, 77
Basterfield, J. G., 215
Baumert, L. D., 238, 251
Baumgartner, J. E., 98
Bender, E. A., 105, 116, 138, 140
Berge, C., 217
Berlekamp, E., 98, 168, 178
Birkhoff, G., 190, 215
Bixby, R., 199, 215
Bollobas, B., 46, 47, 77
Bose, R. C., 219, 240, 245, 247, 251, 252
Bridges, W. G., 14, 19, 20, 21
Brown, T., 198
Brown, W. G., 77, 140
Brualdi, R. A., 20, 203, 215
Bruck, R. H., 10, 233, 251
Brylawski, T., 200, 215, 217
Burr, S., 93, 98

Cameron, P. J., 20
Cates, M., 93, 98
Chernoff, H., 178
Chowla, S., 10, 252
Chvatál, V., 98
Clements, G., 64, 65, 69, 73
Comtet, L., 123, 129, 140
Crapo, H. H., 24, 77, 188, 215, 216
Crawley, P., 216

Daykin, D. E., 58, 65, 67, 76, 77
deBruijn, N. G., 20, 30, 34, 76, 77
Debrunner, H., 216
Dedekind, R., 33
Dembowski, P., 12, 20, 220, 252
Deuber, W., 85, 92, 98
Dilworth, R. P., 4, 20, 28, 29, 34, 77, 205, 216
Doubilet, P., 105, 140
Dowling, T. A., 216
Duffin, R. J., 200, 216
Dumir, V. C., 139, 140
Dushnik, B., 83, 98
Dwass, M., 178

Edelberg, M., 27, 78
Edmonds, J., 18, 20, 188, 216
Egerváry, 4
Ehrhart, E., 140
Erdélyi, A., 129, 140
Erdös, P., 20, 22, 28, 29, 34, 40, 44, 45, 50, 55, 67, 68, 71, 73, 74, 75, 77, 83, 98, 143, 145, 153, 158, 160, 162, 178, 252
Etherington, I. M.H., 129, 140
Euler, L., 116, 219, 245

Fisher, R. A., 221, 252
Foata, D., 140
Folkman, J. H., 85, 98
Ford, L. R., Jr., 4, 20, 79
Freese, R., 28, 77
Frobenius, 4
Fulkerson, D. R., 4, 20, 79, 188, 216

Gallai, 94, 201
Gessel, I., 141

AUTHOR INDEX

Gleason, A. M., 97, 99
Godfrey, J., 65
Goethals, J. M., 20
Goldman, J. R., 77, 105, 140
Goodman, A. W., 20
Gordon, B., 116, 239, 252
Gould, H. W., 140
Graham, R. L., 37, 77, 98, 99, 144, 151, 178
Graver, J. E., 99, 216
Greene, C., 29, 30, 34, 46, 77, 78, 216, 217
Greenwood, R. E., 97, 99
Grunbaum, B., 216
Gupta, H., 139, 140

Hadwiger, H., 216
Hajnal, A., 98
Hall, M., Jr., 8, 12, 20, 140, 233, 248, 252
Hall, P., 4, 20, 39, 78
Hanani, H., 246, 248, 251, 252
Hansel, G., 33, 79
Hardy, G. H., 116, 140
Harper, L. H., 37, 40, 77, 78, 79
Harris, B., 140
Hautus, M. L. J., 141
Henle, M., 105, 140
Herzog, M., 77
Higman, D. E., 252
Hilton, A. J. W., 65, 77
Hindman, N., 92, 93, 98, 99
Hoffman, A. J., 16, 20
Hsieh, W. N., 40, 47, 48, 74, 78

Ingleton, A., 216

Jacobi, 116
Jaeger, F., 214, 217
Jurkat, W. B., 6, 20

Katona, G. O. H., 22, 32, 40, 44, 46, 47, 61, 64, 66, 67, 69, 70, 74, 75, 78
Kelly, D. G., 215, 216
Kelly, J. B., 20, 178
Kelly, L. B., 178
Kelly, L. M., 215

Kennedy, D., 216
Kirkman, Rev. T. P., 218, 252
Klarner, D. A., 111, 113, 140, 141
Klee, V., 216
Kleitman, D. J., 22, 27, 29, 30, 32, 33, 34, 37, 38, 40, 46, 47, 48, 50, 54, 57, 60, 67, 74, 75, 77, 78, 79
Knuth, D. E., 20, 116, 120, 141
Ko, C., 44, 45, 50, 55, 67, 68, 71, 77
König, D., 4, 5, 18
Kramer, E. S., 20, 25
Kreweras, 116
Kruskal, J., 61, 63, 64, 66, 67, 69, 70, 73, 74, 78
Kruyswijk, D. R., 30, 34, 76, 77
Kwiatkowsky, D., 76

Lagrange, 126
Lane, R., 252
Leeb, K., 30, 84, 88, 98, 99
Levine, E., 79
Levitzki, J., 7, 19
Lieb, E. H., 207, 216
Lindström, B., 64, 69, 73, 77, 201
Littlewood, J. E., 74, 79, 116
Lovász, L., 58, 77, 148, 161, 178
Lubell, D., 27, 35, 78, 79
Lucas, T. D., 215, 217

Macauley, F. S., 70, 79
Mac Lane, S., 190, 216
MacMahon, P. A., 116, 120, 141
MacNeish, H. F., 247, 252
MacWilliams, F. J., 20, 213, 252
Magnanti, T. L., 217
Main, R., 216
Mann, H. B., 221, 252
Marcus, M., 8, 20
Markowski, G., 79
Mattson, H. F., 251
McAndrew, M. H., 20
Meschalkin, L. D., 35, 79
Miller, E. W., 83, 98
Mills, W. H., 239, 252
Milner, E. C., 50, 79
Minc, H., 20

AUTHOR INDEX

Mirsky, L., 4, 20
Montgomery, P., 98
Moon, J. W., 141, 151, 153, 178
Moore, E. H., 249, 252

Nešetřil, J., 85, 86, 99
Newman, M., 8, 20

Odlyzko, A., 52
Offord, C., 74, 79
Oxley, J., 214

Parker, E. T., 219, 239, 245, 247, 252
Peele, R., 207
Perfect, H., 20, 21
Peterson, W. W., 168, 178
Pless, V., 251, 252
Polya, G., 111, 141
Popoviciu, T., 114, 141
Posa, L., 166, 178
Pudlák, P., 206

Rado, R., 44, 45, 50, 55, 67, 68, 71, 77, 91, 92, 99, 188, 202, 216
Ramsey, F. P., 80, 82, 84, 95, 99
Raney, G. N., 129, 141
Ray-Chaudhuri, D. K., 252
Read, R., 141
Reid, 199
Rényi, A., 162, 178
Riordan, J., 141
Rödl, V., 85, 86, 99
Rogers, K., 248, 253
Rota, G.-C., 23, 24, 27, 77, 79, 84, 105, 140, 141, 189, 216
Rothschild, B. L., 98, 99, 144
Ryser, H. J., 6, 10, 12, 14, 17, 20, 21, 221, 233, 252, 253

Schoenfeld, L., 140
Schönheim, J., 47, 73, 77, 79
Schröder, E., 129, 141
Schur, I., 81, 91, 97, 99
Schütte, 151
Schützenberger, M.-P., 140
Seidel, J. J., 20, 21

Seymour, 199
Shannon, C., 143, 168
Shirkhande, S. S., 219, 245, 247, 252
Singer, J., 238, 253
Sloane, N. J. A., 20
Spencer, J., 43, 47, 78, 79, 90, 98, 143, 151, 155, 157, 178
Sperner, E., 23, 27, 29, 35, 37, 50, 70, 79
Stanley, R. P., 76, 105, 140, 141, 217
Steiner, J., 218, 253
Stonesifer, J. R., 208, 217
Straus, E. G., 98, 252
Swan, R. G., 21
Sylvester, 116
Szekeres, G., 98
Szele, T., 142, 178
Szemerédi, E., 83, 99

Tamari, D., 129, 141
Tarry, G., 245, 253
Tengbergen, C. A. van E., 30, 34, 76, 77
Thompson, J. G., 20
Tuma, J., 206
Turán, 83
Turyn, R., 251, 253
Tutte, W. T., 21, 178, 188, 199, 214, 217

Ulam, S., 209

Vamos, P., 202
van der Waerden, B. L., 7, 83, 84, 91, 96, 99, 190, 217
van Lint, J. H., 168, 178, 253

Wales, D., 252
Wallis, J., 251, 253
Watson, G. N., 141
Welch, L. R., 239, 252
Weldon, E. J., 168, 178
Welsh, D. J. A., 217
White, N., 203, 217
Whiteman, A. L., 239, 253
Whitman, P., 205
Whitney, H., 179, 180, 190, 195, 199, 217
Whittaker, E. T., 141

Wilson, R. M., 216, 219, 240, 247, 248, 251, 252, 253
Woodall, D. R., 12, 13, 21
Wright, E. M., 116, 140

Yackel, J., 99
Yamamoto, K., 35, 79

Zaslavsky, T. K., 217
Zykov, A. A., 178

SUBJECT INDEX

algebraic function, 123
   irreducible, 123
antichain, 23
   compression of, 66
   order-ideal of, 66
Axiom of Choice, 146

Baire Category Theorem, 82
balanced incomplete block design, 220
    (see also design)
basis exchange axiom, 195
Bell number, 130
bipartite graph, 81
   monochromatic vertex, 81
   r-coloring, 81
   r-Ramsey, 81
biplane, 12
block:
   of incidence structure, 219
   of partition, 130
bracket ring, 203

canonical form, 68, 69
cascade, 63
chain, 23
   maximal, 35
characteristic set, 201
chromatic number, 157
clique, 144
coding problem, 211
coloring, balanced, 32
combinatorial geometry, 182
combinatorial pregeometry, 180
    (see also matroid)
composition, 117
compression, 64, 69
   i-, 73
   reverse, 68

construction:
   direct, 142, 230
   recursive, 142, 244
counting function, 101
critical:
   exponent, 210
   problem, 210

Desargues configuration, 200
design, 220, 221
   Abelian, 232
      multiplier, 233
   group divisible, 226
   Hadamard, 12
   lambda-, 13
   partially balanced, 222
   resolvable, 226
   symmetric, 222, 228
   transversal, 224
   type I lambda-, 13
   (v, k, λ)-, 8
diagonal product, 5
difference set, 231, 238
   Type B, 239
   " B_0, 239
   " GMW, 239
   " H_6, 239
   " O, 239
   " O_0, 239
   " Q, 239
   " S, 238
   " T, 239
   " W_4, 239

entropy, 170
enumeration, 100
   fundamental problem of, 101
error vector, 172

Euler conjecture, 219
Eulerian number, 120

family of sets:
   compressed, 64
   f-sequence, 64

Gaussian coefficients, 26, 207
general position, 181
generating function (s), 100
   chromatic, 103
   Dirichlet, 104
   doubly exponential, 103
   exponential, 102
   Hadamard product of, 114
   in infinitely many variables, 104
   ordinary, 102
   two-variable, 103
geometry (see combinatorial geometry):
   simplicial, 189
girth, 158
Graeco-Latin square, 245
graph:
   bipartitie (see bipartite graph)
   Hamiltonian, 166, 193
   Kuratowski, 199
   planar, 167
   polygon matroid of, 185
   random, 146
   regular, 242
   strongly regular, 242
Grassmannian, 204
group code, 177

Hamming code, 174
Hamming metric, 171, 211

incidence algebra, 106
   reduced, 106
incidence structure (s), 219
   automorphism, 231
   isomorphic, 231
independent set:
   affinely, 181
   of points in a matroid, 180
   of *vertices* in a graph, 144

k-binomial expansion, 62
k-coloring, proper, 213
k-family, 28
k-flow, proper, 214
Kronecker product of matrices, 229

Lagrange inversion formula, 126
Laplace expansion, 202
Latin square, 223
lattice:
   geometric, 185
   of multisets, 25
   of partitions, 27, 204
   of subsets (Boolean algebra), 24, 105, 180
   of subspaces, 26, 105
   point, 184
lines (of incidence structure), 219
line sum, 6
LYM inequality, 35
LYM property, 36

Mac Lane-Steinitz exchange axiom, 185
magic square, 138
   k-component of, 138
      irreducible, 138
   symmetric, 139
matrix:
   doubly stochastic, 7
   formal incidence, 18
   fully indecomposable, 19
   Hadamard, 11, 227
   incidence, 3
   Kronecker product, 229
   line of, 5
   normalized Hadamard, 12
   term rank of, 18
   $(0, 1)$-, 2
matroid, 180
   affine, 181
   algebraic dependence, 190
   binary, 199
   Boolean algebra, 180
   circuits of, 185
   closed sets, 183
   closure, 183

SUBJECT INDEX

connected, 192
contraction, 193
covering relation for, 183
cube, 198
deletion, 193
dependence, 181
dependent sets, 180
direct sum, 191
flats, 183
free, 180
free simplicial, 188
hyperplanes of, 185
independent sets, 180
invariant, 208
isomorphic, 180
isthmus, 195
loop, 183
minor, 195
orthogonal, 195
polygon, 185
presentation, 188
rank function, 187
self-dual, 198
simplicial geometry, 189
support, 197
transversal, 188
unimodular, 199
union, 192
Whitney dual, 195
Möbius function, 205

normalized matching property, 37
n-set, 3

ordered set-system, 43
order ideal, 57
orthogonal, 223
  array, 223
  pair of Latin squares, 223
  pair of rows, 223

packing problem, 212
Pappus configuration, 200
partially ordered set (poset), 105
  antichain of, 23
  binomial, 105
  chain of, 23, 105
    maximal, 35
  interval of, 105
    length of, 105
  k-family of, 28
  locally finite, 105
  normalized matching property, 37
  order-ideal of, 57
  rank function of, 24
  regular, 36
  regular covering by chains, 37
  saturated partition, 29
    symmetric, 30
  Sperner Property, 24
  Whitney Numbers of 2nd kind, 24
partition:
  integer, 101
  set, 27, 130, 204
    blocks of, 130
Pascal Theory, 88
permanent, 5
permutation:
  ascent, 117
  descent, 117
  greater index, 117
  lesser index, 117
plane partition, 121
points of incidence structure, 219
polyomino, 111
  equivalent, 111
P-partition, 116
  order-reversing, 116
  strict, 116
  surjective, 117
  surjective strict, 117
probabilistic method, 142, 146, 147, 153
projective plane, order n, 11

Radon Convexity Theorem, 215
Ramsey:
  category, 86
  Euclidean subset, 94
  numbers, 89
  system of equations, 91
  Theory, 80
    Euclidean, 93

SUBJECT INDEX

random:
  graph, 146
    evolution of, 162
  tournament, 151
rank function, 24, 187
rate of transmission, 170
redundancy, 211
refinement, 27
row sum, 6
r-partition of a set, 130
  blocks of, 130
  convolutional formula for, 132
  type of, 131

semimodularity, 187
simplicial complex, 57, 64
  canonical form of, 64
  f-sequence of, 64
Sperner Property, 24
Steiner triple system, 218
  morphism between, 250
Stirling Numbers of 1st kind, 121
Stirling Numbers of 2nd kind, 27, 121, 207

strict gammoid, 198
syzygy, 203

tactical configuration, 220
threshold function, 162
t-independent subset, 212
tournament, 151
  consistency of, 153
transcendence degree, 190
Tutte-Grothendieck:
  invariant, 208
    characteristic polynomial, 209
    complexity, 209
    subgeometry generating function, 209
  recursion, 208
Tutte polynomial, 208

unimodality, 207

Whitney Numbers of 2nd kind, 24, 207